POWER POLITICS

EQUITY AND ENVIRONMENT IN ELECTRICITY REFORM

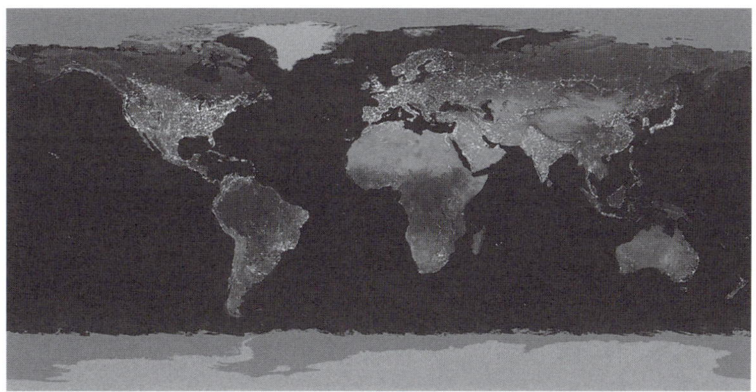

Edited by
NAVROZ K. DUBASH

Contributors:

DANIEL BOUILLE
ALIX CLARK
DIMITAR DOUKOV
HILDA DUBROVSKY
ISHMAEL EDJEKUMHENE
CRESCENCIA MAURER
ELENA PETKOVA
JULIA PHILPOTT
SUDHIR C. RAJAN
AGUS P. SARI
FRANCES SEYMOUR

WORLD RESOURCES INSTITUTE

WASHINGTON, DC

BOB LIVERNASH
EDITOR

HYACINTH BILLINGS
PRODUCTION MANAGER

MAGGIE POWELL
LAYOUT

Each World Resources Institute report represents a timely, scholarly treatment of a subject of public concern. WRI takes responsibility for choosing the study topics and guaranteeing its authors and researchers freedom of inquiry. It also solicits and responds to the guidance of advisory panels and expert reviewers. Unless otherwise stated, however, all the interpretation and findings set forth in WRI publications are those of the authors.

Cover image: Data courtesy Marc Imhoff of NASA GSFC and Christopher Elvidge of NOAA NGDC. Image by Craig Mayhew and Robert Simmon, NASA GSFC.

CONTENTS

BOXES

FIGURES

TABLES

ACKNOWLEDGMENTS

This study was a collective effort by several people. The contributors to this volume – from the World Resources Institute (WRI) and from partner organizations — devoted considerable time and energy to the study. They brought not only commitment to and ideas for a strong final product, but also belief in the need for a more sustainable global electricity future. I have been privileged for the opportunity to work with them.

Within WRI, my colleagues in the International Financial Flows and the Environment (IFFE) project, Frances Seymour and Crescencia Maurer, provided core intellectual input and support in conceptualizing and implementing this study, in addition to contributing chapters to the volume. Two additional colleagues in the IFFE project, first Lily Donge, and then Linda Shaffer Bollert, managed the research and production process, while also providing research assistance. Without their work, this document would not have been completed. I deeply appreciate their efforts.

Several other colleagues contributed their time and effort in completing this study. WRI colleagues Crescencia Maurer, Lily Donge and John Coyle, with assistance from Gretchen Hoff, Fritz Kahrl, and Mehr Latif compiled the text boxes on country cases. Yu-Mi Mun from the University of Delaware contributed a box on South Korea. Mara Angeloni, a visiting fellow from the Italian Ministry of Environment and Territory, and Amy Cassara compiled and analyzed data on pollutant emissions. Kate Ling and Ken Ogino assisted with research, production and outreach during their summer internships at WRI.

This study draws on reflection and input of many people from governments, donor agencies, civil society organizations and industry. I thank our sources who generously contributed their time and knowledge during interviews and in correspondence. Additional thanks to colleagues who participated in a workshop held at WRI in October 2000 to review preliminary papers.

The quality of this work has benefited considerably from experts who reviewed the entire manuscript: Sanjeev Ahluwalia, William Ascher, Walt Patterson, and Ian Tellum. In addition, I appreciate detailed review of portions of the report by: Armarquay Armar, Daniel Azpiazu, John Besant-Jones, Abeeku Brew-Hammond, Istvan Dobozi, Elsa du Toit, Michael Eckhardt, Paul Galen, Rumen Gechev, David Hawes, Gilberto Januzzi, Appiah Korang, Jaime Millan, Lulin Radulov, Amulya K.N. Reddy, Martín Rodríguez Pardina, Anil Terway, Nurina Widagdo, and Harald Winkler. In addition, I thank WRI colleagues who reviewed an early draft of the volume: Elizabeth Cook, Paul Faeth, Nancy Kete, Wendy Vanasselt, and Patricia Zurita. The reviewers' comments and suggestions have considerably strengthened the report. The authors retain full responsibility for any remaining errors of fact or interpretation.

Hyacinth Billings, Robert Livernash, and Maggie Powell worked rapidly and diligently to edit, produce, and publish the report. Johnathan Kool produced graphics for the report. I thank them and additional members of WRI's communications staff for their assistance with this publication.

This study, and the work of the International Financial Flows and the Environment project, is made possible by grants from the Charles Stewart Mott Foundation, the Wallace Global Fund, the Spencer T. and Ann W. Olin Foundation, the Netherlands Ministry of Foreign Affairs, and the U.S. Environmental Protection Agency. I am grateful to them for their support.

NAVROZ K. DUBASH

FOREWORD

Over the last decade the hitherto staid, stable world of electricity supply has become tumultuous. Across the globe, rich and poor countries alike have begun a march toward restructuring their electricity sectors around market competition. These reforms have typically been the province of technocrats, who have designed reforms to meet narrow economic and technical objectives.

In *Power Politics: Equity and Environment in Electricity Reform*, Navroz Dubash and his colleagues from the World Resources Institute (WRI) and around the world show how electricity reform is, at root, an issue of sustainable development. Electricity reform represents an opportunity to focus attention on the 1.7 billion of the world's poor without access to electricity. It could also be an opportunity to align investor incentives along a trajectory toward a clean energy future, one that reduces emissions of greenhouse gases while promoting development and supporting livelihoods. The concern is not solely one of a missed opportunity. Inappropriately done, electricity reform could hinder progress toward a more socially and environmentally sustainable energy future.

Drawing on six case studies from the developing world and economies in transition, the contributors to this volume examine whether and how the process of electricity reform can support rather than hinder sustainable development. Starting from the premise that political support is critical if electricity reform is to support sustainable development, the authors examine the political drivers and interests at the heart of this process.

Instead of sustainable development, they find that financial concerns and donor conditions have driven electricity reform. Managed by closed political processes and dominated by technocrats and donor consultants, environmental considerations play almost no role in a re-envisioned electricity sector. Social concerns are given more importance, but only to the extent that reforms affect politically powerful groups. Donor agencies, such as the World Bank, have been central to stimulating reform, and will be important actors helping to determine the future of the sector.

In order for a restructured sector to contribute to sustainable development, governments and donors will need to factor concerns of sustainability into reform design early, and back them up with political commitment. Civil society groups have a key role to play by laying the political groundwork for this agenda, and by holding decisionmakers and regulators accountable for their decisions. Finally, through their choice of investments and support for good governance in the sector, private investors can contribute toward a more sustainable electricity sector. The report provides additional recommendations for each of these groups.

This report complements other work at WRI that addresses how institutions and governance can influence energy and climate trajectories, including: *The Climate of Export Credit Agencies* and *Public Finance Restructuring for Sustainable Development in Emerging Market Economies.* In addition, this study adds to WRI's body of work on the effects of national and international policy change on global financial flows including: *The Right Conditions: The World Bank, Structural Adjustment, and Forest Policy Reform;* and *Will International Investment Rules Obstruct Climate Protection Policies?*

Support for this study and for the International Financial Flows and the Environment Project has come from the C.S. Mott Foundation, the Wallace Global Fund, the Spencer T. and Ann W. Olin Foundation, the Netherlands Ministry of Foreign Affairs, and the U.S. Environmental Protection Agency. I am pleased to express our appreciation for their generosity and foresight.

JONATHAN LASH
PRESIDENT
WORLD RESOURCES INSTITUTE

EXECUTIVE SUMMARY

During the 1990s, the conventional wisdom about the electricity sector—public ownership and integrated utilities—was challenged by a new model of private ownership and unbundled utilities. Debates about the viability, applicability, and feasibility of market-led electricity reforms continue today. Nonetheless, at the turn of the new century, countries around the world are taking tentative steps toward this new approach.

These shifts in the electricity sector have not occurred in isolation. The new model is part of a broader thrust toward the promotion of markets, a growing role for private capital, and global economic integration. These themes place electricity sector reforms squarely within larger processes of economic globalization and the debates about its merits and costs.

Electricity sector reforms and the financial flows they attract have serious implications—potentially both positive and negative—for long-term sustainable development goals. A sector designed to ensure access to electricity for all could bring considerable social benefits, including opportunities for education, better health and nutrition, and entrepreneurship. A sector designed with environmental considerations in mind could significantly mitigate the build-up of global and local pollutants. Failure to address these social and environmental concerns—collectively "public benefits"—could undermine progress toward sustainable development.

Decisions made now about the institutional structure and functioning of the electricity sector will shape social and environmental outcomes for decades to come. Whether market-led or not, reforms will best support sustainable development outcomes when they are explicitly designed to do so.

The central question for this study is: *How can the process of reforming the electricity sector support rather than hinder promotion of sustainable development outcomes?* We approach this question by examining the process and politics of reform in six countries in the developing world and economies in transition—Argentina, Bulgaria, Ghana, India, Indonesia, and South Africa. These countries were selected to ensure a mix across early and late reformers, large and small countries, and to provide a geographic spread. To answer the central question, each country study asks:

- What were the drivers of reform in the electricity sector?
- What political interests were at stake in reform of the sector, and how did they shape the reform process?
- What role did the World Bank and other international donor agencies play in electricity sector reforms?
- How and by whom were social and environmental concerns addressed in the process of designing electricity reforms, and with what outcomes?

Each country study was conducted as a collaborative exercise between the World Resources Institute and a research partner from the country studied. Specific issues in a small number of additional countries were briefly examined to supplement the main case studies. Our methods were semi-structured interviews with key informants from government agencies, civil society, the private sector, and international agencies—all conducted on a not-for-attribution basis to encourage candor. This information was supplemented by official government and donor agency reports, other secondary materials, and media reports.

ELECTRICITY SECTOR REFORM AND A SUSTAINABLE DEVELOPMENT AGENDA

Reform of the electricity sector is on the agenda in much of the developing world and in transition economies. Diminished barriers to private capital flows, technological change in power generation technologies, and ambitious early experiments with institutional restructuring in Chile and the United Kingdom have stimulated reform efforts around the world. In developing and transition economies, a World Bank policy of conditioning loans on institutional restructuring provided a further impetus to reform. By 1998, of a sample of 115 developing countries, 33 percent had passed new electricity laws, 29 percent had established an independent regulator, and 40 percent had allowed the entry of privately owned independent power producers (IPPs) (Bacon, 1999).

The approach to reform will determine whether it supports or undermines sustainable development. Electricity restructuring will influence important social concerns such as access to price, quality of service, and labor impacts. In a restructured electricity market, price signals and a profit motive alone will be insufficient to ensure that social goals in the sector are met. *(See Box.)*

Electricity reform will also shape the future environmental profile of the sector. Market incentives for economic efficiency will likely result in greater environmental efficiency in the short run. However, reforms may not help realize a clean energy future in the absence of explicit planning mechanisms that factor in environmental benefits and costs. Electricity restructuring also provides a rare opportunity to spur the transition to a "micropower" future based on small-scale distributed generation. To do so, reform designers will have to be attentive to the environmental implications of economic regulatory decisions in order to provide a level playing field and ensure that reforms do not reduce opportunities for end-use energy efficiency. *(See Box.)*

CASE STUDY FINDINGS

The six case studies suggest that, with the exception of South Africa, there has been little political commitment to promoting sustainable development through electricity sector reforms.

Argentina: Reforms in Argentina were stimulated by a severe macroeconomic crisis in the late 1980s. Facing hyperinflation, a heavy debt burden, and declining quality of public services, Argentina's reform program was intended to reduce the government's role in providing key services, including electricity. The reforms were designed by a small group of politically powerful bureaucrats—supported by multilateral agencies such as the World Bank—with little scope for broader debate. The reforms did lead to improved quality of service in urban areas and increases in system efficiency. However, they also undermined incentives to increase energy efficiency, limited expansion of electricity to isolated rural populations, placed a disproportionate burden on low-income consumers, and failed to effectively manage expansion of the transmission system. A second generation of reforms in the late 1990s has attempted to address some of these concerns.

India: In India, concerns over the financial state of the sector dominated reform design. In 1991, the government provided incentives for electricity generation to stave off a balance-of-payments crisis. The effort to attract private capital not only failed to increase capacity as planned, but also locked the sector into adverse financial and institutional arrangements. The World Bank played a central role in initiating a second stage of state-level reforms beginning in 1996 to address the fundamental problem of inadequate revenue flow in the sector.

<table><tr><td>BOX</td><td>

HOW DOES ELECTRICITY REFORM AFFECT SOCIAL AND ENVIRONMENTAL CONCERNS?
</td></tr></table>

Social

Access: In a restructured electricity market, profit alone is often an insufficient driver for expanding access to relatively unprofitable rural customers and the urban poor. Incentive schemes, subsidies, or regulatory mandates may be required.

Price: Electricity reforms are typically associated with pressures to limit subsidies and enhance tariff collection. While these changes make for a better functioning sector, the resultant price increases can also cause social hardships and spur political opposition to reforms. A mitigation strategy can address these costs.

Quality: Competition in restructured markets may increase the reliability, choice, and responsiveness of electricity service providers, but is not guaranteed to do so absent appropriate regulation and oversight.

Labor: Public sector electric utilities face job cuts as a result of reforms. This retrenchment will bring social costs. Opposition from labor interests can be a political deterrent to reforms and will have to be addressed and mitigated.

Environmental

Technology/Fuel choice: The choice of technology and fuel used to generate electricity has environmental impact. The market structure put in place by reforms can affect technology choice by changing the relative attractiveness of capital-cost intensive technologies versus those based on high running costs. In addition, the existence and basis of a planning framework for electricity will determine whether environmental considerations factor into a long-term vision for the sector.

Regulatory decisions: Economic regulatory decisions often also have environmental outcomes. Regulators can influence how level the playing field is for different technologies. They can also implement a strategic vision for the sector. Regulators must have the mandate and training necessary to play these roles.

Incentives for efficiency: Electricity reforms that enforce financial discipline should contribute to greater efficiency of supply, with environmental gains. However, reforms can introduce additional transaction costs, and obscure price and other signals to customers, raising obstacles to end-use efficiency improvements. Conversely, competition could spur retailers to market end-use efficiency services.

State-level reforms have produced mixed results at best. Privatization efforts have been fraught with difficulty. Where utilities have been privatized, the change has not produced expected gains. Efforts at promoting public benefits—such as energy efficiency at the state level and incentives for renewable energy sources—have been relatively few and have suffered from a lack of political commitment.

Indonesia: Early efforts at attracting private capital for electricity generation in Indonesia in the mid-1990s occurred under a shadow of corruption. These efforts also invited World Bank disapproval,

reversing long-standing donor support for Indonesia's power sector. The result was the construction of costly excess generation capacity, which colored future reform efforts. The 1997 Asian financial crisis spurred an attempt at broader reform as part of an IMF-led economic adjustment strategy. The post-crisis reform effort was accompanied by a consultation process personally led by the Minister of Energy and Mines. This process was stalled by political upheavals, unresolved issues with IPPs, and the political challenge of raising tariffs. Social equity—in particular concerns over tariffs—have been at the forefront of reform debates, while environmental concerns have scarcely influenced reform design.

Bulgaria: Reforms in Bulgaria were initiated by an IMF stabilization program in 1997 following a period of financial crisis, but the reform program was shaped by national political currents. Government-led reforms have been driven in large part by a determination to become an energy exporter, despite evidence that this is not a viable strategy—a position that was only reversed with a change in government in 2001. Despite Bulgaria's environmental obligations under the Kyoto Protocol and its candidacy for European Union membership, environmental concerns did not play a role in shaping reforms. After an initial focus on financial issues and prices, donors have actively promoted attention to the considerable gains to be achieved from encouraging energy efficiency in the economy as part of a reform strategy. Under a new government, a shift in political focus has improved the prospects for this approach.

Ghana: Reforms in Ghana were driven by a shortage of financing for much-needed capacity expansion in 1995; sector reform was a condition of World Bank lending for new capacity. But the Ghanaian government set aside the Bank's recommendation for limited reforms and took the initiative to develop a more extensive design. An important political actor in this process was the large and powerful Volta River Authority, which initially feared its position in the sector would be threatened by reforms. Although expansion of access to electricity is a significant issue in Ghana, the government failed to integrate existing electrification efforts with institutional reforms. While there was little explicit focus on environmental issues in the course of reform design, measures to promote energy efficiency and provide incentives to renewable energy sources were added to reform efforts.

South Africa: Reforms in South Africa are driven by a broader national agenda to restructure state-owned enterprises, initiated in the mid-1990s. Reform in the electricity sector began in earnest in the late 1990s. While financial considerations are important in South Africa, reforms have not been spurred by an immediate short-term financial crisis, either in the sector or in the economy at large. As a result, the national government has exercised considerable control over reforms, and has framed them around social issues such as access to energy and black economic empowerment. The existing public utility, Eskom, has been an important political actor in discussions about whether this agenda is better served by the existing system or by a restructured sector. In addition, reforms in South Africa have provided scope for broader consultation and debate, a process in which donor agencies have played a restricted, information-provision role.

A comparison across the case studies suggests several common themes:

Electricity reforms are driven by economic and financial concerns, and by donor conditionalities.

Reforms in Argentina, Indonesia, and Bulgaria were undertaken in an environment of macroeconomic crisis. In India, Indonesia, Bulgaria, and Ghana, donor conditions were the immediate reason for undertaking reforms. As a consequence, reform design was often driven by an immediate need to attract capital—a trend reinforced by donor agencies. However, efforts to attract capital, particularly through IPPs, have caused more problems than they have solved. In India and Indonesia, IPP entry has been accompanied by allegations of corruption and undermined the financial and institutional health of the sector. In Argentina, the urgent need for capital led to privatization at reduced prices. While reforming countries are criticized for not providing sufficient incentives to attract foreign capital, it is not clear whether such incentives are politically viable and socially desirable. Structuring reforms mainly to attract finance may not be a sustainable long-term strategy for the sector. Moreover, the focus on financial issues crowds out attention to public benefits.

Closed political processes and politically powerful groups constrain attention to sustainable development objectives.

To a large extent, reforms were designed by government bureaucrats and their consultants in the energy and finance ministries, to the exclusion of other voices. In Argentina, for example, reforms were designed and implemented with great speed by a small group of technocrats. Even within governments, the cases show little evidence of involvement by environment and rural development ministries in the design stage of reform. Despite a vibrant civil society, the cases do not provide instances of participation or influence by nongovernmental organizations (NGOs) in policy design, even though several NGOs have been active in this area. South Africa—with a more open reform design process, greater engagement by a range of ministries, and more participation by outside experts—is an exception.

In all the cases, tariff increases and restructuring have proved to be the single biggest sticking point to electricity reform and have been greeted by popular uprisings in Argentina, India, Indonesia, Ghana, and South Africa. Powerful political constituencies have also been obstacles to reform. In Ghana and South Africa, existing public utilities initially argued for their continued viability as integrated public entities. Faced with the possibility of socially destabilizing labor retrenchment, labor unions have been a political force against reform. However, in both Argentina and Bolivia, unions won a share in the equity of privatized state enterprises demonstrating the possibility of political compromise.

The case studies do not conclusively demonstrate that an open process is preferable to the quick and stealthy approach to reforms. The threat remains that open reform processes could be politically captured by narrow interests. However, there are indications that an open process is the better alternative. To be politically sustainable, the public must believe that reforms will lead to demonstrable benefits—an outcome that is better supported by a transparent process. An exclusive process is also prey to being subverted and used for narrow ends by the new wielders of authority, as was arguably the case with the experience of IPPs in Asia. An open process would provide checks on such abuses of power.

Donor agencies have initiated reforms and advocated attention to environmental concerns, but have been hampered by past reputation and a perception of favoring private interests.

Donor agencies have been central to cutting through a domestic political morass to initiate reforms. In India, it took World Bank intervention for governments at the state level to agree to seriously examine the need for new institutional and financial arrangements. While this initial firmness may have been necessary, a continued heavy hand in steering reforms undermined domestic ownership, with negative consequences. For example, donors sought to expand the role for the private sector and establish the conditions for profit making in Ghana and India, when it was not clear that the regulatory environment was sufficiently developed to support those changes.

At the same time, donor agencies have often taken the lead in preparing studies and undertaking projects related to the environmental dimensions of electricity reform. World Bank studies on the environmental impact of restructuring have been influential in shaping policy in Bulgaria, as have efforts by the Danish government to promote renewable energy in Ghana. Often, however, these efforts have been late, too restricted in scope, and not backed by adequate political signals.

Moreover, donor agencies' efforts to provide assistance have been hampered by a reputational burden built over a decade or more of controversial structural adjustment policies, which the public associated with economic hardship and undue promotion of private sector interests. This reputation has been worsened by the industrialized countries' efforts to promote the interests of their own corporations. Such was the case in Indonesia, where one arm of the U.S. government sought to promote a large U.S.-funded IPP, even as an advisor supported by its aid agency, USAID, cautioned against the project.

To be effective, public benefits need to be factored into reform design early and backed by political commitment.

For reform designers, ensuring a financially viable sector was the most relevant definition of public benefits. Social and environmental concerns were matters to be grafted onto reforms at a later stage. However, the Argentina experience—where reforms led to subsidy removal and tariffs that were skewed against low-income groups—suggests that a *laissez faire* approach does not automatically support social objectives and can undermine equity in outcomes. Since technical, political, and institutional decisions made during reforms constrain future choices, it is hard to retrofit the sector to address public benefits.

For example, IPPs in India and Indonesia locked those countries into large generation plants. This undermined efforts at energy efficiency and committed utilities to buy electricity at uncompetitive prices. In another example, regulators' mandates, priorities, and skills were established in the early stages of reform. Without attention to sustainable development goals in the inception process, it will be an uphill battle to re-direct regulators' attention from short-term concerns to longer-term social and environmental concerns.

These longer-term concerns merit attention. In several countries shifting to a decentralized, market approach has contributed to the absence of a broad vision for the sector. In Argentina, this absence undermined the integrity of the transmission system. In India, the central government has belatedly attempted to forge a broad vision to guide state-level reforms. In Bulgaria, a vision for the future was initially built on an unviable export strategy. Most significantly, pressing social and environmental concerns have not been integrated into reforms. In India and Ghana, the process of institutional reform was not coordinated with ongoing, and ineffective, electrification programs. By contrast, in South Africa reforms have been closely associated with a political commitment to expand access to electricity. In Bulgaria, international environmental commitments have not played a role in electricity reform, despite the sector's considerable environmental footprint. Without a broad vision and political support, the case studies suggest that public benefits are prey to political whims and shifting trends in donor assistance.

RECOMMENDATIONS: TOWARD A PROGRESSIVE POLITICS OF ELECTRICITY SECTOR REFORM

Integrating environmental and social benefits into electricity sector reforms in developing and transition economies will continue to be a daunting challenge. Not only are reforms technically complex, but the combination of macroeconomic crisis, entrenched political interests, and centrality of costs often crowd out attention to environmental and social factors. However, the country studies do offer insights into how reforms are currently shaped, and therefore into how attention to concerns of equity and sustainability can be reinserted into the reform process.

1. Frame reforms around the goals to be achieved in the sector.

A narrow focus on institutional restructuring driven by financial concerns is too restrictive to accommodate a public benefits agenda. To build a framework that includes such an agenda requires an articulation of the services that a reformed sector is intended to provide and the means by which it should do so. While donor agencies often play a central role in initiating reform, they must step back during the process of defining goals to allow a nationally-driven vision of reform to emerge.

2. Structure finance around reform goals, rather than reform goals around finance.

Reform processes have catered to a need to attract private capital. Since sustainable development may not always be aligned with short-term profit motives, reform processes must move beyond the imperative of attracting capital. While this may seem a far-fetched notion in capital-constrained developing countries, the time may now be opportune to change the terms on which private capital enters a country. Efforts to attract capital through risk mitigation and tariff increases have not won popular backing, and as a result have not been politically sustainable. A broader vision of reform and a public consensus supporting that vision could lower these risks. Private capital may be willing to accept more realistic financial returns, if they are combined with less risk. Political legitimacy in a reform program, tied to some innovation in mechanisms for raising finance, may be a more promising route than tailoring reforms to short-term profit horizons.

3. Support reform processes with a system of sound governance.

An open-ended framing of reforms will reflect public concerns only if it is supported by a robust process of debate and discussion. Hence, a third imperative is to embed debate over electricity sector reforms in a sound process of decisionmaking guided by transparency, openness, and participation. Such an approach is more likely to provide the political space for articulation of a range of public concerns than have the closed processes prevalent thus far. It is also more likely to build public consensus in support of reforms, making for a more politically sustainable process.

4. Build political strategies to support attention to a public benefits agenda.

It is important that public benefits advocates strengthen political coalitions supporting sustainable development and counter those favoring parochial interests. In particular, the case studies suggest that social concerns carry far more political weight in a national context than do either local or international environmental issues. Efforts to exploit links between social and environmental agendas would likely be a useful political approach.

By focusing on financial health, reforms in the electricity sector have excluded a range of broader concerns also relevant to the public interest. In this study, we have examined the social and environmental concerns at stake in these reforms. We have found that not only are they inadequately addressed, but that socially and environmentally undesirable trajectories can be locked-in through technological, institutional, and financial decisions that constrain future choices. Consequently, social and environmental benefits need to be internalized early in reform decisionmaking.

To do so, the process by which reform goals are defined and reform decisionmaking must change to embrace a more consensus-driven design of reforms. More complex processes bring with them greater risks of capture by special interests and failure due to a cacophony of voices. Yet exclusive reforms of the electricity sector have not incorporated the breadth of interests that deserve a voice and have not yet shown themselves to be sustainable—financially, socially, or environmentally. This study has suggested several reasons to believe that a modified approach guided by a vision of a socially and environmentally sustainable electricity future may yield a more satisfying outcome.

REFERENCE

Bacon, Robert. 1999. *Global Energy Sector Reform in Developing Countries: A Scorecard.* Report 219/99. Washington, D.C.: ESMAP.

1

INTRODUCTION

Navroz K. Dubash

Over a three-year period in the early 1990s, a small team of technocrats transformed the electricity sector in Argentina. Responding to a burgeoning foreign debt and a growing crisis of management in the sector, Argentina terminated any direct government role in electricity supply. Instead, they transformed electricity into a commodity to be bought and sold on an open market. The intended goal was to reap efficiency gains, and the means was to minimize government interference with the market's hidden hand. A decade later, South Africa also embarked on electricity sector reform. In the midst of a program of post-apartheid reconstruction and development, South Africans grappled with how to retain a place for the sector as an instrument of poverty alleviation consistent with environmental sustainability, even while re-making it to capture market efficiencies.

The Argentina and South Africa experiences with power sector reform bracket a decade of change in the sector. During this period, the conventional wisdom favoring public ownership and operation of the electricity sector was challenged by a new paradigm of market competition for electricity. Whether and how countries should follow this approach is still a subject of debate, in part because of the considerable problems encountered by the state of California following sector reforms there. Still, the approach has won many adherents. By 1998, a survey of 115 developing countries found that 73 had taken at least minimal steps down the road to market-oriented reforms in the electricity sector (Bacon, 1999). In part due to these changes, $187 billion was invested in energy and electricity projects in developing and transition economies between 1990 and 1999 (World Bank, 2000).

Electricity sector reforms and the financial flows they generate carry considerable implications—potentially both positive and negative—for long-term sustainable development goals. *(See Box 1.1.)* This report is motivated by the concern that decisions made now about the institutional structure and functioning of the electricity sector will shape patterns of development for decades to come. Our approach is informed by the view that electricity reforms—market-led or not—can best support socially and environmentally progressive outcomes when they are explicitly designed to do so. Consequently, the central question for this study is:

How can the process of reforming the electricity sector support rather than hinder promotion of sustainable development outcomes?

BOX 1.1 | **ELECTRICITY AND SUSTAINABLE DEVELOPMENT CHALLENGES**

- Fifty-six percent of the world's rural population does not have access to electricity.*
- Electricity generation accounts for 38 percent of worldwide CO_2 emissions. **

* World Energy Assessment, 2000, p. 374.

** Computed by WRI from IEA data.

In posing this question, we examined the process by which electricity sector reforms are initiated, designed, and implemented in six country studies—Argentina, Bulgaria, Ghana, India, Indonesia, and South Africa. To understand whether reforms are likely to contribute to sustainable development by explicit inclusion of a public benefits agenda in national reform processes, we examined the national politics that shape reform in each country. Since national reforms are influenced by larger global trends, we also explored how national circumstances are shaped by international intervention, particularly by international aid agencies.

Reforms in the electricity sector need not follow a single prescription.

As the examples of Argentina and South Africa illustrate, not all countries have followed the same path toward power sector reform. In Argentina, reforms were dictated by a rigid application of market principles. In South Africa, reform efforts were embedded within a broader debate over economic and social empowerment. These different approaches suggest that reforms in the electricity sector need not follow a single prescription. By understanding the forces that shape reform, this study can suggest ways toward a more progressive politics of electricity reform.

THE LINK BETWEEN ELECTRICITY AND SUSTAINABLE DEVELOPMENT

The electricity sector has long been an integral part of the engine of economic growth. It is also a central component of sustainable development.[1] High-quality energy, which includes access to electricity services, can be a powerful force for development. Access to electricity supports improvements in health, education, and opportunities for entrepreneurship. Yet it is estimated that 1.7 billion people lack access to electricity (World Energy Assessment, 2000). The effect of sectoral reforms on incentives to

BOX 1.2 | **HOW DOES ELECTRICITY REFORM AFFECT SOCIAL BENEFITS?**

Access: In a restructured electricity market, profit alone is often an insufficient driver for expanding access to electricity to relatively unprofitable rural customers and the urban poor. Incentive schemes, subsidies, or regulatory mandates may be required.

Price: Electricity reforms are typically associated with pressures to limit subsidies and enhance collection of tariffs. While these changes make for a better functioning sector, the resultant price increases can also cause social hardships and spur political opposition to reforms. A mitigation strategy can address these costs.

Quality: Competition in restructured markets may increase the reliability, choice, and responsiveness of electricity service providers, but are not guaranteed to do so in the absence of appropriate regulation and oversight.

Labor: Public sector electric utilities face job cuts as a result of reforms. This retrenchment will bring substantial social costs. Opposition from labor interests can be a political deterrent to reforms and will have to be addressed and mitigated.

provide broad access to electricity services—and on the price at which these services are available—can be a significant determinant of human development. *(See Box 1.2.)*

The electricity sector is a significant consumer of fossil fuels. In addition to the environmental impacts of fossil fuel extraction, the sector is responsible for a substantial share of local and global pollutants. Decisions made now—as the sector is reformed and electricity markets restructured—will create both incentives and disincentives for large or small-scale generation, fossil fuels or renewable energy technologies, efficient or inefficient supply and use of energy, and centralized or decentralized generation sources. *(See Box 1.3.)* To be sustainable, sector reforms must incorporate attention to social and environmental benefits, referred to in this study as "public benefits."

Incorporating public benefits need not follow past approaches, which as the case studies show have sometimes been misguided. For example, electricity subsidies in India and Bulgaria both encouraged wasteful consumption and did not benefit poor populations. Moreover, the ensuing financial shortfalls undermined technical performance and worsened environmental outcomes.

Decisions made now about the electricity sector will shape patterns of development for decades.

The challenge for the future is twofold. First, as discussed in Chapter 2, it is to develop electricity sector reform policies and approaches that promote sustainable development while supporting a well-functioning electricity sector. Second, as discussed in the rest of this study, it is to ensure that sustainable development remains part of the political calculus that drives reforms.

THE GLOBAL CONTEXT

Reforms in the electricity sectors of individual countries have occurred in the context of global economic integration. The broad contours of this process include political, economic, financial, technological, and institutional transformations. These trends are briefly spelled out here and discussed in more detail in Chapter 2.

The globalization of the 1980s and '90s presumed a growing faith in the market as an instrument of economic coordination. This shift was accompanied by an expanding role for private corporations and a

| BOX 1.3 | **HOW DOES ELECTRICITY REFORM AFFECT ENVIRONMENTAL BENEFITS?** |

Technology/Fuel choice: The choice of technology and fuel used to generate electricity has environmental impact. The market structure put in place by reforms can affect technology choice by shifting the relative attractiveness of capital-cost intensive technologies versus those based on high running costs. In addition, the existence and basis of a planning framework for electricity will determine whether environmental considerations factor into a long-term vision for the sector.

Regulatory decisions: Economic regulatory decisions often also have environmental outcomes. Regulatory decisions influence how level the playing field is for different technologies. They can also implement a strategic vision for the sector. Regulators must have the mandate and training to play these roles.

Incentives for efficiency: Electricity reforms that enforce financial discipline should contribute to greater efficiency of supply, with environmental gains. However, reforms can introduce additional transactions costs and obscure price and other signals to customers, raising obstacles to end-use efficiency improvements. Conversely, competition could spur retailers to distinguish themselves by marketing end-use efficiency services.

corresponding questioning and renegotiation of the appropriate role of the state in economic activity. In developing countries, a turn toward markets and away from state-led activity was promoted by two decades of World Bank structural adjustment policies, which were intended to increase resource-use efficiency by enlarging the scope for private sector activity (Jayarajah and Branson, 1995). In

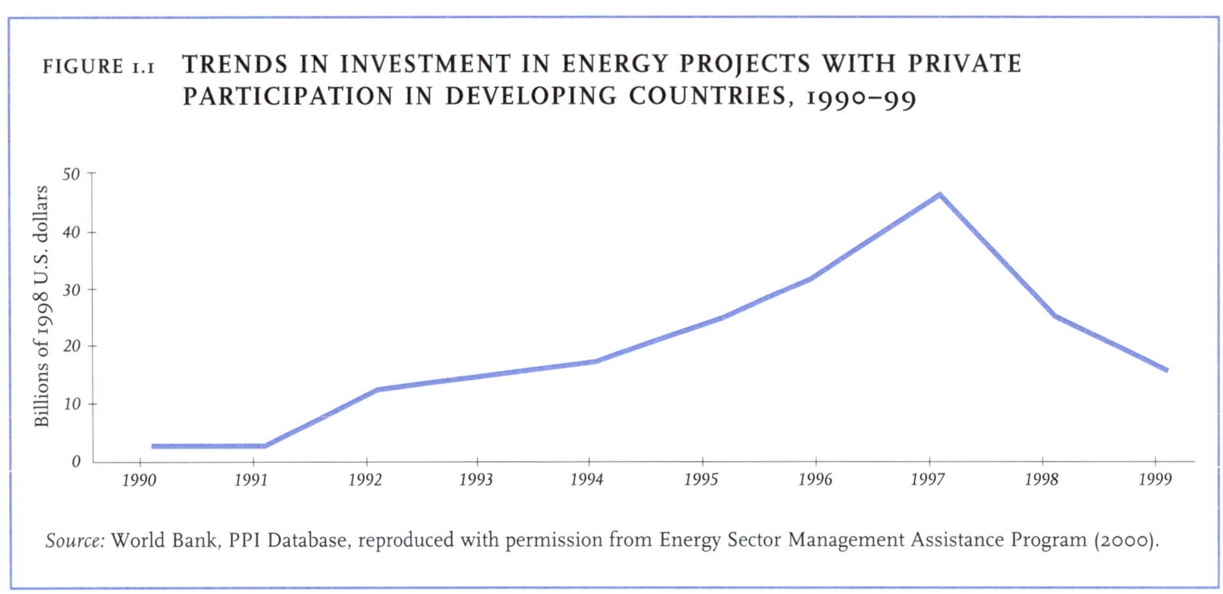

Source: World Bank, PPI Database, reproduced with permission from Energy Sector Management Assistance Program (2000).

particular, the structural adjustment agenda included privatization of state-owned corporations, such as electric utilities, as part of a regime of fiscal responsibility.

This transition facilitated a greater role for international private capital in the economies of developing countries. As Figure 1.1 shows for the energy sector as a whole, during the 1990s private capital flows increased considerably—albeit with a downturn following the Asian financial crisis of 1997. The entry of private finance influenced the institutional form of the sector. In order to contain risks, private financiers sought to invest in discrete projects rather than around an entire power system, where risks were harder to measure and manage.

The trend toward financing smaller, discrete units rather than an integrated whole was further supported by technological changes in the electricity sector. The development of small and cheap gas turbines reversed the decades-long trend toward growing economies of scale in electricity generation (Patterson, 1999). Moreover, the development of information technology and computing power vastly increased the capacity to monitor, control, and measure electricity usage and flows (Graham and Marvin, 1995). Together, these technological shifts

undermined the natural monopoly characteristic of the power sector and challenged the centralized nature of public utilities.

The precedent for institutional reform of the electricity sector was set by developments in Chile and the United Kingdom (Bacon, 1995; Rosenzweig and Voll, 1997). In both countries in the 1980s, public monopolies responsible for generating and providing electricity were restructured. The component parts were sold to the private sector and placed under a regulatory framework intended to encourage competition. In both cases, the transformation to private ownership and competition was driven more by ideological considerations than by evidence of the benefits of restructuring. While there has been considerable debate about the effectiveness and replicability of this approach, the demonstration effect of the Chile and U.K. experience was undoubtedly considerable. Following the experience in these two countries, restructuring to encourage competition in the electricity sector became a viable policy option. In particular, urged by the World Bank and other donor agencies, developing countries and transition economies have considered both privatization and restructuring as policy options to address problems of cash shortages, capacity shortfalls, and poor management.

Viewed in this context, reforming the electricity sector is about far more than adopting alternative technical and institutional models. Current patterns of electricity reform are being integrated into larger processes of globalization—notably the predisposition toward markets, the growing role of private capital, and efforts to weave the sector into the fabric of international economic integration. Electricity sector reform debates are part of a broader dialogue about how to organize economic life and ensure the public interest in a globalizing environment. As with other dimensions of globalization, debates over the electricity sector are marked by a polarization of views.

Believers in economic integration—typically economists, regulatory consultants, and some staff of donor agencies—argue for reforms aimed at unfettered markets for electricity. In the transformation of electricity provision into a business venture, they see the potential for greater efficiency and dynamism, with a corresponding public benefit in the form of lower prices and better service. From this perspective, the electricity sector has suffered from too much interference by the state and from too many misguided, if well-meaning, efforts at steering the sector toward social and environmental gains.

The opposite view—often held by representatives of nongovernmental organizations (NGO's), some developing country utility managers, and some staff in international organizations—is that a transition to privatization and competition will lead to an unchecked search for profit and a betrayal of the public interest. According to this camp, the obituary for the public utility model has been written in unseemly haste. From this perspective, electricity is a public service that should be guided by broader public objectives such as social and environmental goals.

Each of these admittedly caricatured views is problematic, as we discuss here in brief, and explain in greater detail in Chapter 2. The belief that unfettered markets will automatically generate public benefits is dubious on both environmental and social grounds. Left to their own devices, markets are ill-equipped to address equity considerations in access to electricity or prices. They fail to internalize environmental impacts and, therefore are likely to provide less than optimal environmental and social benefits. On the other hand, a stubborn adherence to the past ignores the currently dismal technical and financial state of the sector in many countries, and is likely to be swamped by the new realities of financial globalization and rapid technological change. A future electricity sector—whether public or private—that resists adaptation and flexibility is just as unlikely to serve the public interest as one that looks to the market alone to provide the right signals.

Markets are ill-equipped to address equity considerations in access to electricity or prices.

In this report, we suggest that markets are socially and politically constructed. If they are to contribute to the public interest, they must be explicitly designed to do so (Evans, 2001). Since markets do not exist in a political vacuum, whether or not reforms serve the public interest depends first on how effectively these concerns emerge from political processes. This is not to trivialize the intellectual and ideological debates about questions of public versus private ownership, and bureaucratic versus market coordination, but to emphasize that reform debates too often focus on false dichotomies. A more useful approach is to understand how the political decisionmaking process shapes electricity sector reform.

THE POLITICAL ECONOMY OF REFORM

Under what circumstances does reform in the electricity sector incorporate attention to public benefits and promote sustainable development? Reforms are nominally under the control of governments, whether at the federal or at subsidiary levels. But they rarely are initiated, designed, and implemented by the state alone in a simple sequence. Reform processes are shaped by the interaction of formal rules, norms of behavior, and sociopolitical environments. Consequently, reforms both require and lead to a rearrangement of the relationships among states, the private sector, and civil society (Brinkerhoff and Crosby, 2002).

Governments are seldom homogeneous and coherent entities. Different ministries typically have divergent interests, which figure heavily in shaping reform agendas. In addition, politicians may not share the same interests as bureaucrats, and different branches of government—the executive and the legislative, in particular—often disagree. In the electricity sector, public utility officials have much at stake in electricity reforms, and may themselves be significant players in reform dialogues.

In addition, reforming governments face tremendous pressures from below. These pressures may be applied by well-organized beneficiary coalitions with claims on public resources (Waterbury, 1992). The country studies that follow illustrate, for example, the political clout of large farmers in India who have grown used to subsidized electricity for agriculture, and ruling elites in Indonesia who used the sector as a vehicle for graft. Actors such as these are well mobilized and in a position to voice opposition to reforms. By contrast, potential beneficiaries of proactive environmental and social policies are often diffuse and poorly organized—as for example rural populations that are unconnected to the grid. Promotion of public benefits may thus require searching for new coalitions and new players— consumer organizations, environmentalists, pro-poor advocates, municipalities, and some private corporations.

The private sector stands to gain enormously from a reform agenda that promotes greater private participation in the power sector. Some may try to influence the process of reform and seek concessions from governments. In this context, even if liberalization in the electricity sector leads to state exit from

the direct business of supplying electricity, ensuring governmental functions nonetheless requires a strong and capable state role (Brinkerhoff and Crosby, 2002). Other private sector firms may emerge in reaction to this opportunity by, for example, developing opportunities for investment in energy efficiency or providing access to electricity through distributed generation technologies. This group may be potentially important members of a coalition to promote attention to public benefits.

Electricity reforms present civil society groups with an opportunity, but also a challenge. Over the last two decades, civil society groups have focused on centralized, often state-owned electricity bureaucracies. They have targeted socially and environmentally destructive projects, often in alliance with international NGOs that amplify their reach to donor agencies and international financiers (Keck and Sikkink, 1998; Hildyard and Mansley, 2001). The larger number of actors and increasing complexity of decisionmaking in a restructured electricity sector challenges NGOs to re-think their strategy. Some view restructuring as an opportunity to replace staid bureaucracies with a dynamic, decentralized sector that will serve sustainable development goals (Hirsh and Serchuk, 1999). Others are more circumspect about the nature of this opportunity, and see little reason for hope that a restructured sector will promote a more sustainable energy future (PRAYAS, 1999; Mun, 2000; Tellam, 2000; Dixit, Wagle, and Sant, 2001). A strong belief among NGOs is that open, transparent, and effective governance will be a key ingredient in realizing public benefits in a restructured sector. Consequently, demands for a more participatory and open decisionmaking process are likely to be an important part of the national politics of electricity reforms.

While pressures from below are imposing, pressures from above can be just as sizeable. International lenders, called in to bail out crisis-ridden economies, have significant leverage over the trajectory of reforms in borrower countries. Correspondingly, the space, time, and flexibility to shape reforms available to domestic actors, both governmental and nongovernmental, may shrink. At the same time, there is an active debate about how much leverage international donors really have over domestic processes (Kahler, 1992), in what context aid conditionality is appropriately deployed (World Bank, 1998), and whether and how donor leverage can be made to address social and environmental aims beyond traditional growth objectives (Nelson and Eglinton, 1993). This study assesses the role of international donors in the reform process and asks whether and how they can contribute to the promotion of public benefits in the course of electricity sector reforms.

Whether and how different actors and interests participate in shaping the electricity sector depends on the governance structure under which reforms are carried out. There is a broad divide between two opposing viewpoints on the appropriate governance of reform processes (Williamson, 1994; Rodrik, 1996). One view holds that economic reforms must be carried out by a strong executive, unhampered by the need to consult or seek consensus, in order to stop vested interests from obstructing a reform agenda. From this perspective, while reforms may be rational for society as a whole, myopia on the part of the general public and a collective irrationality nonetheless can obstruct reform. The opposing position argues the central importance of forging a social consensus around reform. Consultation not only provides the promise of improving policy, but, by addressing the concerns of the general population, it raises the probability of continued support for a reform program, and supports democratic institutions. By contrast, an autocratic approach, even if tied to good economics, can make for undesirable politics by undermining democratic institutions.

The distinction between these two positions is blurred somewhat by noting that while the initiation of reforms may require a firm and autonomous executive with a relatively free hand, consolidation of reforms may rest in building consensus (Rodrik, 1996). If effective implementation requires not only good design, but agreement and cooperation from relevant agencies and non-state stakeholders, then the case for consensus building is further strengthened (Brinkerhoff and Crosby, 2002).

By permitting trade-offs to be explicit, and providing avenues for a broader range of interests and perspectives to be brought to the table, it appears likely that the social consensus approach is likely to be more favorable to promotion of sustainable development. The counter-argument, however, is that if reforms are brought to a halt in a futile search for consensus, then even the benefits of limited reforms will be lost. The relative benefits of the two approaches is an issue considered in the country chapters.

FRAMEWORK OF THE STUDY AND RESEARCH METHODS

Chapter 2 sets the stage for the country studies. It provides the global context for electricity sector reform by describing both the forces leading to an institutional shift in the sector and the historical process by which this shift occurred. Next, it describes and briefly summarizes debates over the prevalent model that emerge from the Chile and U.K. experience. It then examines how this model spread around the world, with an emphasis on the growing role of private sector financing. Finally, it makes the argument that a *laissez faire* approach to market-led reform in the electricity sector is unlikely to fully serve the public interest.

To further our understanding of electricity sector reforms as shaped by political processes, this study is organized around six country case studies (Chapters 3-9) from the developing world and economies in transition. The case studies—Argentina, India, Indonesia, Bulgaria, Ghana, and South Africa—were selected to provide a mix across early and late reformers, large and small countries, and also to provide a geographic spread. They are presented in the chapters that follow organized from earliest reformer to most recent. In addition, the country studies are complemented by brief examinations of reform processes in other countries, the results of which are described throughout the chapters as text boxes. The central concern of this study is how electricity sector reform processes can promote sustainable development through attention to public benefits. Each country study asks:

- What were the drivers of reform in the electricity sector?
- What political interests were at stake in the sector and in its reform, and how did they shape the reform process?
- What role did the World Bank and other international donor agencies play in electricity sector reforms?
- How and by whom were public benefits concerns addressed in the process of designing reforms, and with what outcomes?

The country studies were conducted as a collaborative exercise between the World Resources Institute and a research partner from each country. Our primary methods were semi-structured interviews with key informants from government agencies, civil society, the private sector, and international agency staff—all conducted on a not-for-attribution basis to encourage candor. Accordingly, interviews cited in the studies are referenced only by the affiliation of the interviewee. This information was supplemented by official government and donor agency reports, other secondary materials, and media reports.

In the conclusion, we examine the results of the country studies in comparative context, and draw some implications about steps toward a more progressive politics of electricity sector reform.

NOTES

1. We follow the Brundtland Commission (World Commission on Environment and Development, 1997) in defining sustainable development as "development that meets the needs of the present, without compromising the ability of future generations to meet their needs." We also follow Lele (1991) in further specifying that sustainable development at a minimum includes a desire to limit environmental degradation, while allowing for basic needs and economic growth, and seeking to accomplish these outcomes in a participatory fashion. Finally, we acknowledge the significant conceptual weaknesses of the term, but note its political utility as an umbrella concept.

REFERENCES

Bacon, R.W. 1995. "Privatization and Reform in the Global Electricity Supply Industry." *Annual Review of Energy and Environment* 20:119-143.

Bacon, Robert. 1999. "A Scorecard for Energy Reform in Developing Countries." Public Policy for the Private Sector Note No. 175. Washington, D.C.: World Bank.

Brinkerhoff, Derick W., and Benjamin L. Crosby. 2002. *Managing Policy Reform: Concepts and Tools for Decision-Makers in Developing and Transition Economies.* Bloomfield, CT: Kumarian Press.

Dixit, Shantanu, Subodh Wagle, and Girish Sant. 2001. "The Real Challenge in Power Sector Restructuring: Instilling Public Control through Transparency, Accountability, and Public Participation." *Energy for Sustainable Development* V (3):95-101.

Energy Sector Management Assistance Program. 2000. *Energy Services for the World's Poor.* Washington D.C.: World Bank.

Evans, Peter. 2001. "Looking for Agents of Urban Livability in a Globalized Political Economy." In *Livable Cities: Urban Struggles for Livelihood and Sustainability*, edited by P. Evans. Berkeley, CA: UC Press.

Graham, Stephen, and Simon Marvin. 1995. "More than Ducts and Wires: Post-Fordism, Cities and Utility Networks." In *Managing Cities: The New Urban Context*, edited by P. Heal, S. Cameron, S. Davoudi, S. Graham and A. Madani-Pour. New York: John Wiley & Sons.

Hildyard, Nicholas, and Mark Mansley. 2001. *The Campaigners' Guide to Financial Markets.* Dorset, UK: The Corner House.

Hirsh, Richard F., and Adam H. Serchuk. 1999. "Power Switch: Will the Restructured Electric Utility System Help the Environment?" *Environment* 41 (September):4-29.

Jayarajah, Carl, and William Branson. 1995. *Structural and Sectoral Adjustment: World Bank Experience, 1980-92.* Washington D.C.: World Bank.

Kahler, Miles. 1992. "External Influence, Conditionality, and the Politics of Adjustment." In *The Politics of Economic Adjustment*, edited by R. R. Kaufman. Princeton: Princeton University Press.

Keck, Margaret and Kathryn Sikkink. 1998. *Activists Beyond Borders.* Ithaca, NY: Cornell University Press.

Lele, Sharachchandra M. 1991. "Sustainable Development: A Critical Review." *World Development* 19 (6):607-621.

Mun, Yu-Mi. 2000. "Electric Utility Restructuring and its Implications for Ecological Sustainability and Democratic Governance: A Preliminary Review." Unpublished paper. Center for Energy and Environmental Policy, University of Delaware.

Nelson, Joan M., and Stephanie J. Eglinton. 1993. *Global Goals, Contentious Means: Issues of Multiple Aid Conditionality.* Washington, D.C.: Overseas Development Council.

Patterson, Walt. 1999. *Transforming Electricity.* London: Earthscan.

PRAYAS. 1999. "Regaining Rationality through Democratisation: A Critical Review of MDBs Power Sector Activities in India." Unpublished paper. Pune, India.

Rodrik, Dani. 1996. "Understanding Economic Policy Reform." *Journal of Economic Literature* XXXIV (March):9-41.

Rosenzweig, Michael B., and Sarah P. Voll. 1997. "Sequencing Power Sector Privatization." Unpublished paper. Washington D.C.: National Economic Research Associates.

Tellam, Ian, ed. 2000. *Fuel for Change: The World Bank Energy Policy, Rhetoric vs. Reality.* London: Zed Books.

Waterbury, John. 1992. "The Heart of the Matter? Public Enterprises and the Adjustment Process." In *The Politics of Economic Adjustment*, edited by S. Haggard and R. R. Kaufman. Princeton, NJ: Princeton University Press.

Williamson, John, ed. 1994. *The Political Economy of Policy Reform.* Washington, D.C.: Institute for International Economics.

World Bank. 1998. *Assessing Aid: What Works, What Doesn't, and Why.* Washington, D.C.: World Bank.

World Commission on Environment and Development. 1997. *Our Common Future.* Oxford: Oxford University Press.

World Energy Assessment. 2000. *World Energy Assessment: Energy and the Challenge of Sustainability.* Edited by United Nations Development Programme, United Nations Department of Economic and Social Affairs, and World Energy Council. New York: United Nations Development Programme.

2

THE CHANGING GLOBAL CONTEXT FOR ELECTRICITY REFORM

Navroz K. Dubash

To a significant extent, national electricity sector reform initiatives were shaped in response to global trends in development ideology, financing, and technological change. In this chapter, we review these trends and examine their implications for electricity reforms. Based on this discussion we explore a debate over whether preserving and enhancing the public interest in the sector will occur automatically as a result of reforms, or whether it must be designed into reform processes.

ORGANIZATION OF THE ELECTRICITY SECTOR: FROM SOCIAL COMPACT TO ECONOMIC EFFICIENCY

Until the early 1990s, governments either owned the electricity sector or controlled the sector through regulation. Electricity was considered a textbook natural monopoly (Teplitz-Sembitzky, 1990; Hunt and Shuttleworth, 1996). Governments, it was thought, were best able to mobilize the large amounts of capital necessary to develop the sector and bear the long time horizons for recovery of costs. Particularly in developing countries, government leadership in the development and use of electricity was part of a broader "social compact" (World Bank, 1988).

In the early 1990s, however, this conventional wisdom came under siege. With astonishing rapidity, there was a revolution in thinking about the structure of the industry. In the industrialized world, the promise of the public, vertically integrated, centralized power system was largely realized—people had reliable, affordable power. The problems were those of a mature system and technologies. By the 1970s, however, industrialized countries no longer benefited from the smoothly rising demand curves of the past, undermining the predictable sources of income on which utilities had relied, and affecting the returns from new power projects (Rosenzweig and Voll, 1997). When small, cheap gas turbines became commercially viable, the trends in scale economies that had dominated the industry until this point were dramatically reversed (Hunt and Shuttleworth, 1996).[1] Costs and risk in the sector had increased due to a rising environmental consciousness, a corresponding increase in regulations, and burdensome investment in large power plants, particularly high-capital-cost nuclear units (Patterson, 1999). In short, a virtuous cycle of increasing consumption, growing interconnection, and lower costs driven by scale economies was replaced by a vicious cycle of increasing costs, diminishing productivity, and deteriorating economic performance (Oliviera, 1997).

Countries began to act on these changed realities. In the 1970s, the United States allowed independent power producers to sell electricity to utilities. This was a shift of considerable significance. It demonstrated that independent generators could be integrated into a grid system, and began the unraveling of the conventional wisdom that the utility was a natural monopoly (Hirsh, 1999). In the late 1980s,

Chile and the United Kingdom took reform a step further by re-making their sectors around the objective of promoting competition.[2] This was a radical idea.[3] Before these countries initiated reform, there was near unanimity that transaction costs in the sector and the technical requirements of electricity made competition nearly impossible.[4]

There was a rush to anoint a new conventional wisdom—competition in the power sector was not only possible, but inevitable.

The rapid growth and declining costs of communication and information technologies facilitated the development of new control techniques consistent with decentralization, which facilitated competition (Graham and Marvin, 1995; International Energy Agency, 1999). In a few short years, there was a rush to anoint a new conventional wisdom—competition in the power sector was not only possible, but inevitable. This new model represented a shift from a "social compact" to the pursuit of economic efficiency.

A COMPLEX NEW MODEL FOR THE ELECTRICITY SECTOR

The emerging model in the electricity sector is focused primarily on two dimensions: (1) changes in management practices, which may or may not involve changes in ownership from government to the private sector (*See Box 2.1.*); and (2) restructuring for competition, which is a process of separating or "unbundling" vertically integrated utilities to progressively introduce competition into the system (*See Box 2.2.*).[5] In positioning themselves with respect to these issues, countries have a wide range of choices. However, implementing reforms is by no means as simple as picking locations along these two axes. Putting into practice a competitive model for electricity requires developing rules for markets, contractual arrangements, tariff regulation, and a myriad of other details.

BOX 2.1 | **STAGES IN CHANGE OF OWNERSHIP AND MANAGEMENT**

Commercialization: A government surrenders detailed control over a state-owned enterprise with the purpose of promoting operation in keeping with commercial principles. This is a change in practice rather than a change in organization form.

Corporatization: A government formally and legally relinquishes control and management of a state-owned enterprise to establish a corporation. It may still set overall objectives and subject the corporation to regulatory oversight.

Privatization: A government sells a corporation to private owners. The private corporation is able to tap the capital markets, and is subject to the discipline of those markets. The private company may be subject to regulatory oversight.

Source: Hunt, Sally and Graham Shuttleworth. 1996. *Competition and Choice in Electricity.* New York: John Wiley and Sons.

In addition to technical details, there are conceptual challenges to implementing this model. Privatization and restructuring, while theoretically distinct processes, are linked in practice. Privatization alone, in the absence of competitive market structure, will do little to promote competition (Oliviera and MacKerron, 1992). However, the introduction of truly competitive market structures will limit profits, at least in comparison to firms with market power. Ironically, the successful introduction of competition could mute private sector interest in the sector and actually undercut successful privatization (Bacon, 1995). Conversely, if privatization occurs before restructuring, private owners have a strong incentive to ensure that subsequent reform efforts do not undermine their ability to capture monopoly rents through, for example, effective regulation. This may open the door to corruption and other means of influencing the restructuring process. Privatization and competition, then, while apparently complementary, pose considerable challenges of sequencing and implementation (Newbery, 1995; Besant-Jones, 1996).

To further complicate the picture, competition need not be accompanied by privatization. For example, Norway has successfully introduced competition between different state-owned entities, some owned by the central government and others by municipalities (Wolak, no date; Magnus, 1997).

The benefits and viability of implementing a market model, either in whole or in part, continue to be hotly debated. Advocates of market-based reforms suggest that reforms provide incentives for cost savings and productivity enhancement (Joskow, 1998). These incentives operate by providing appropriate price signals to consumers to allocate resources appropriately, by unleashing the profit motive to provide an incentive for efficient use of inputs, and by encouraging cost reductions through competition (Bacon and Jones, 2001). They point to the experience of early movers among the industrialized countries, such as the United Kingdom, to suggest that these goals have been realized (Littlechild, 2000). However, the U.K. experience also shows that competition is at least as important as privatization in providing incentives for efficiency, and that the costs of system coordination are greater under the market system than in an integrated utility (Newbery and Green, 1996).

The benefits and viability of implementing a market model continue to be hotly debated.

Critics focus on the weaknesses of the market competition model (Watts, 2001). For example, the cost of capital for the sector is likely to be higher in an unregulated market (to reflect higher risks) than under either public ownership or regulation based on a stable rate of return. In a capital-intensive sector like electricity, the higher cost of capital can result in both higher average cost and greater price variability. In addition, it is difficult to defend against market power in this sector, both because electricity cannot be stored (giving generators opportunities to exploit market power), and because there is little elasticity of demand for electricity in the short run. Finally, planning for the transmission system becomes a challenge in a privatized sector where investment decisions may be a trade secret, and where choices

<table>
<tr><td>BOX 2.3</td><td># THE CALIFORNIA ELECTRICITY CRISIS</td></tr>
</table>

The Plan

In 1993, California was beginning to emerge from a prolonged recession. Electricity rates were 50 percent higher than the U.S. national average, owing—among a host of other factors—to the high costs of California's nuclear power plants and expensive long-term power contracts. Lawmakers feared that the high price of energy was driving industry out of the state. Deregulation was intended to reduce the price of electricity by introducing market competition.

In 1996, after two years of discussion and debate, the California Legislature unanimously adopted a plan for deregulation. To improve prospects for competition, the plan offered significant incentives to California's three largest utilities to separate their generation, transmission, and distribution functions into component parts.

The utilities quickly sold off their oil and natural gas power plants; their market share in electricity generation fell from 81 percent in 1996 to 46 percent in 1999. They also transferred control of electricity transmission to an independent system operator and established a power exchange facility to facilitate electricity sales between generators and utilities. The system was designed to allow residential customers to choose their electric service provider and thus inject retail competition into the energy sector.

Significantly, the plan allowed the utilities to recover their "stranded costs"—capital costs in plants that were potentially uneconomic in the new competitive generation market—by freezing consumer retail rates at 1996 levels for up to four years or until the costs were recovered, whichever came first. Since it was assumed that wholesale rates would remain well below retail rates, the plan would allow the utilities to lock in a sure profit. This was the carrot that made them willing to give up their near-monopoly on generation and transmission of electricity. The end result was a complex, delicately balanced system whose design represented a series of political compromises made between legislators, interest groups, and the utilities.

Crisis

In 2000, a combination of factors—some relating to supply, others to demand—combined to plunge California's power sector into crisis. On the supply side, a drought in the Pacific Northwest reduced the amount of hydroelectric power available for export to California in the spring months of 2000. Natural gas prices quadrupled in the same time period. Some have argued that the state's strict emissions standards limited the output of currently existing plants. Environmentalists counter that California's environmental regulations played no part in precipitating the crisis but that lack of supply was manufactured by generators exercising market power. All agree, however, that unbundling had failed to dilute market power.

The demand for electricity in California surged by 20 percent from 1997 to 2000, driven in part by a boom in the technology sector. Yet because retail prices were fixed by law, there was no way to signal to consumers that their higher demand was driving up the price of electricity. Moreover, in the mid-1990s the utilities reduced funding for what had been successful energy efficiency programs which would otherwise have lowered demand.

Exacerbating these problems was a lack of flexibility in the plan. Assumptions that supply reserves were adequate, that there would be robust competition, and that consumers would adjust use in response to market signals, all proved to be flawed. Although there were official public hearings, a significant amount of the drafting process took place in late-night, closed-door, horse-trading sessions between legislators, utilities, and environmental and other interest groups. The end result was a bill designed to satisfy each group. According to some insiders, by appeasing major interest groups, the consensus-building process may have deflected serious scrutiny of the bill. Finally, because the stakeholder consultation process led to the program being institutionalized in law—rather than in regulations, which are more flexible—policymakers had fewer options when dealing with a fast-changing and growing crisis in 2000 and 2001.

In mid-May 2000, as excess capacity dwindled, wholesale energy prices began to rise. By June, the utilities were buying power at $120 per megawatt-hour and then selling to retail customers at the fixed price of $65 per megawatt-hour. Two utilities declared their stranded costs recovered and pleaded with the California Public Utilities Commission (CPUC) to lift the retail price restrictions. It refused, arguing that consumer rate hikes were politically impossible. The crisis was exacerbated by market manipulation; suppliers withheld energy at peak-hours in an attempt to drive up prices. The average number of megawatts off-line from November 2000 to May 2001 increased by 267 percent compared to the same period a year earlier. California's utilities began losing a great deal of money very quickly.

By mid-December 2000, the utilities were paying $400 per megawatt-hour for power on the wholesale market, reselling it for $65 per megawatt-hour, and losing roughly $50 million per day. By mid-January 2001, two utilities were insolvent, with one declaring bankruptcy in April. The power exchange shut down on January 31, but the governor instructed the California Department of Water Resources (CDWR) to buy power to meet the utilities' short positions. Through May 2001, CDWR had spent about $8 billion doing so.

Only in the spring of 2001 did a combination of conservation efforts, price caps, and the willingness of a previously reluctant government to sign long-term contracts with energy providers bring wholesale prices down to earth. The state of California cut its energy use by an astonishing 8 percent during the month of February 2001. A reduction in demand helped to nudge wholesale prices downward and to reduce the incentive for energy generators to take capacity off-line.

By July 1, 2001, electricity prices were back to the levels of early May 2000. The California government has effectively taken control of the electricity sector, and is now saddled with billions of dollars in long-term contracts that require it to buy power at prices significantly above current market rates. This will almost certainly ensure rate hikes and tax increases for

California's citizens for the foreseeable future. This outcome is a far cry from the promises made in 1996.

The California story offers several possible lessons for reforming countries. The complexity of the reform challenge is daunting, suggesting the need for keeping reform simple. In particular, California's experience illustrates how pernicious the problem of market power can be; unbundling California's utilities only transferred the problem from one cartel to another. With regard to the process of reform, California did make a serious attempt at broad consultation, which allowed a broad range of interests to be represented. That the end result was unsuccessful suggests that while stakeholder engagement may be a necessary component of any reform plan, it is not in itself sufficient to ensure success. Finally, conservation and energy efficiency efforts were remarkably effective and a major factor in bringing the energy crisis to a close.

Sources:

The Economist. 2001. "A State of Gloom." (January 20).

Gamson, David. 2001. "The California Conundrum: Rates, Reliability and the Environment Under Electric Restructuring in California." Paper presented at "Energy Restructuring and the Environment" Workshop in Tokyo, Japan. (October 22).

Harvey, Hal, Bentham Paulos, and Eric Heitz. 2001. "California and the Energy Crisis: Diagnosis and Cure." San Francisco, CA: Energy Foundation.

Joskow, Paul L. 2001. "California's Energy Crisis." Working Paper 8442. Cambridge, Massachusetts: National Bureau of Economic Research. (August).

Nieves, Evelyn. 2001. "In California, Solar's Time to Shine: Alternate Power Systems and Conservation Draw New Converts." *The New York Times.* (May 3).

Reddy, Amulya K.N. 2001. "California Energy Crisis and its Lessons for Power Sector Reform in India." *Economic and Political Weekly.* (May 5-11).

World Bank Energy and Mining Sector Board. 2001. "The California Experience with Power Sector Reform: Lessons for Developing Countries." Washington, D.C.: World Bank. Online at: http://www.worldbank.org/html/fpd/energy/pdfs/e_calexpo400.pdf (May 29, 2002).

between transmission investments may considerably benefit one generator over another.

Others do not take issue with the model itself, but are cautious about the pragmatics of implementation in developing countries. In small developing countries, there is a trade-off between dividing up generators sufficiently to ensure competition, and ensuring economies of scale in generation (Bacon, 1995; Besant-Jones, 1996). Moreover, the starting conditions in the sector are an important factor in determining the viability of the private competition market. For developing countries, whether retail prices are above or below costs, or capacity generation is adequate to meet demand, the extent to which the population is fully served by electricity access and the existence of credible regulatory institutions to restrict the emergence of private monopolies will all determine decisions on how to proceed with reform of the sector (World Bank, 2001). Finally, while a market reform model provides incentives for efficient operation at the level of the firm, it does not by itself provide incentives for the balanced development of the sector, and would particularly neglect unprofitable customers or those unconnected to the grid (Reddy, 2001).

The debate has acquired renewed intensity in the wake of California's disastrous experience in the summer of 2000 following an extremely ambitious, though arguably flawed, reform program. *(See Box 2.3 on previous page.)* The collapse of the Enron Corporation in the United States—the company most closely and visibly associated with electricity markets—has also raised questions about the relative merits of deregulation and government control.[6]

It is beyond the scope of this study to conclusively consider these issues. However, based on the brief review conducted here, it seems true that the model described here may indeed generate gains, but there is considerable uncertainty about how best to define and implement reforms in particular cases. In particular, how reforms can best be tailored to the varying conditions in different countries along both privatization and restructuring dimensions, and whether and how the greater complexity of the

sector can be adequately governed, remain open questions.

ADOPTION OF THE MODEL: THE STRUGGLE FOR FINANCIAL RESOURCES

Developing countries and economies in transition have faced some of the same issues as industrialized countries, notably the emergence of new generation and information technologies and, to some extent, rising costs due to increased regulation. But in other ways the problems have been quite different. These problems are outlined in Chapters 3-9 in greater detail.

As summarized in the *World Energy Assessment*, state-owned monopolies in many countries have allowed subsidies to proliferate, demonstrated a bias in favor of large and visible projects, been prey to bad management, and have placed a strain on government budgets (World Energy Assessment, 2000). For example, subsidized power was often used to propel forward key sectors of the economy. This approach, while effective, also created powerful constituencies for the continuation of such policies, as was the case with farmers in India. In other countries, nontransparent accounting undermined financial discipline in the energy sector, as for example in Bulgaria through the use of coal subsidies for electricity and electricity subsidies for district heating. The sector was often used for narrow political ends, resulting in a weakening of public institutions in the sector across many developing countries (Teplitz-Sembitzky, 1990). This observation was reinforced by a survey of 300 World Bank-funded projects from 1965 to 1983, which showed a progressive deterioration in the performance of developing country utilities over time (World Bank, 1988). In short, there is considerable evidence that the electricity sector faces significant problems in much of the developing world and in economies in transition.

The type and magnitude of problems are not similar all over the world. For example, in contrast to many other developing countries, the sector in Argentina and South Africa is well-developed and functions relatively well. In Bulgaria, and in much of Central and Eastern Europe, the sector is well-developed but plagued by inefficiencies (Chandler, 2000). For example, energy intensity (the amount of energy required to produce a unit of GDP) in this region is approximately twice as high as in other industrialized countries (Tellam, 2000). In other countries, such as India and Ghana, the promise of the public utility model has failed to materialize. Most African countries provide electricity to less than 20 percent of their population (Bhagavan, 1999). Those who do have access to electricity are often inadequately served.

Despite these differences across countries, a market-based reform approach has fast taken root in many developing countries. While few have taken all the steps described in Boxes 2.1 and 2.2, several countries have undertaken some combination of steps toward a market in the electricity sector. *(See Table 2.1.)* For example, 44 percent of 115 countries surveyed in 1998 had converted their state utility into a corporation, and 40 percent had allowed the entry of private producers in generation (Independent Power Producers). A smaller number of countries had privatized either generation (21 percent) or distribution (18 percent). Nonetheless, taking into account the dramatic nature of institutional change required, these proportions represent a considerable shift over less than one decade.

What explains this relatively rapid adoption of the model in the developing world? To a significant extent, the answer lies in the search for finances for the energy sector in developing and transition economies. Traditionally, much of the developing world has relied heavily on public development finance, and particularly the World Bank, to finance their investments in the sector. By the 1990s, international public financial institutions were increasingly reluctant to continue funding public utilities that were trapped in a cycle of low revenues and declining quality. This led to a steep decline in World Bank funding for investment projects in the electricity sector. *(See Figure 2.1.)*

In addition, continuing a decade of "structural adjustment" in borrower countries, the World Bank and the International Monetary Fund sought to expand the role of the private sector in the development process. In 1993, a World Bank policy paper made reform in the electricity sector an explicit condition of continued lending for the sector (World Bank, 1993). The central thrust of the new policy was to encourage borrower nations to restructure their sectors and open them to greater private participation. Toward this end, the World Bank increased lending for policy reform. *(See Box 2.4.)* This shift was not limited to the World Bank, but is echoed in a 1994 energy sector policy paper produced by the Asian Development Bank (Asian Development Bank, 1994).

Obtaining private finance for the electricity sector was no easy task. The institutional framework for

---

TABLE 2.1 | **COUNTRIES TAKING KEY REFORM STEPS IN THE POWER SUBSECTOR, 1998**
(sample of 115)

Corporatize	Law	Establish Regulator	Independent Power Producers	Restructure	Privatize Generation	Privatize Distribution
51	38	33	46	40	24	21
(44%)	(33%)	(29%)	(40%)	(35%)	(21%)	(18%)

Source: Bacon, Robert. 1999. *Global Energy Sector Reform in Developing Countries: A Scorecard.* Report 219/99. Washington, D.C.: ESMAP.

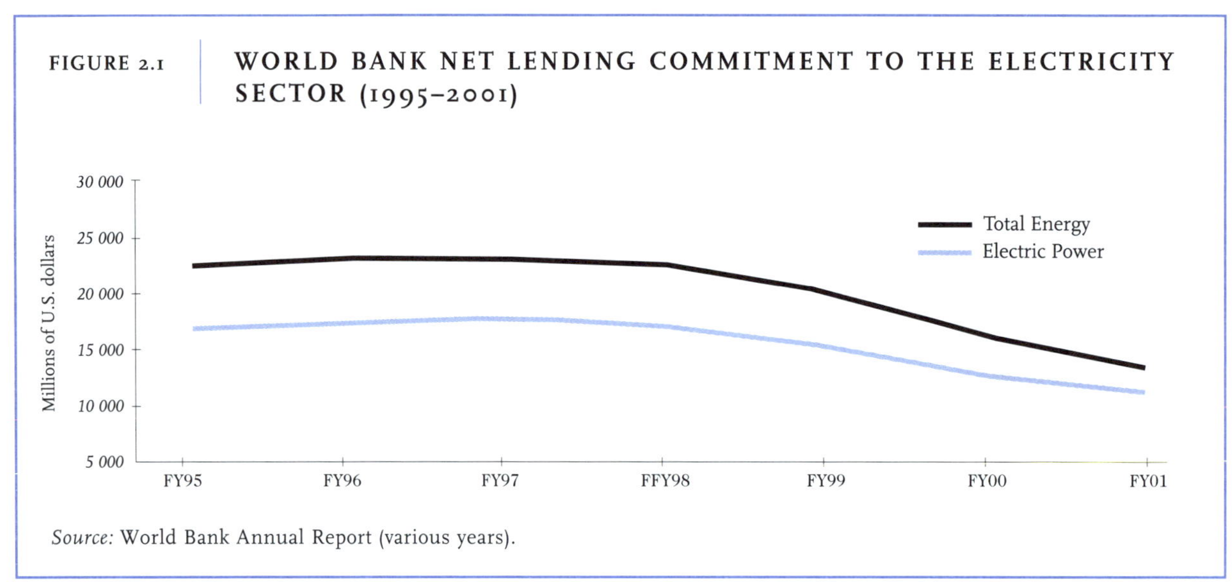

Source: World Bank Annual Report (various years).

private investment in the sector did not exist. Like the experience in the United States, the United Kingdom, and Chile, developing countries and economies in transition had to pass new laws and establish new institutions to attract capital. In addition, under the public utility model, the sector was organized as an interconnected network. This structure did not lend itself to discrete investments with well-defined profiles of risk and return to private capital. Instead, dependence on private capital exerted a pressure to divide the sector into discrete components (Balu, 1997). Finally, the poor state of the sector in many potential recipient countries did not promise either reasonable expectation of profit or manageable low risk. Hence, borrowing countries were in a bind: to attract capital, the sector had to be in good health, and in order to ensure good health, they needed capital.

BOX 2.4 | SHIFTS IN WORLD BANK POLICY IN THE ELECTRICITY SECTOR

From 1978 to 1993, World Bank lending in the electricity sector supported state-owned monopoly power utilities to:

- provide power service on the basis of least-cost development programs;
- strengthen the sector's institutions and improve their efficiency;
- increase local resource mobilization; and
- improve access to electricity by disadvantaged population groups.

In 1993, the World Bank made reform an explicit condition of continued lending for electricity. The strategy called on borrower countries to:

- establish transparent regulatory processes;
- commercialize and corporatize the power sector;
- allow for importation of power services in some cases; and
- encourage private investment in the sector.

Sources:

Covarrubias, Alvaro J. 1996. *Lending for Electric Power in Sub-Saharan Africa.* Washington, D.C.: World Bank.

World Bank. 1993. *The World Bank's Role in the Electric Power Sector.* Washington, D.C.: World Bank.

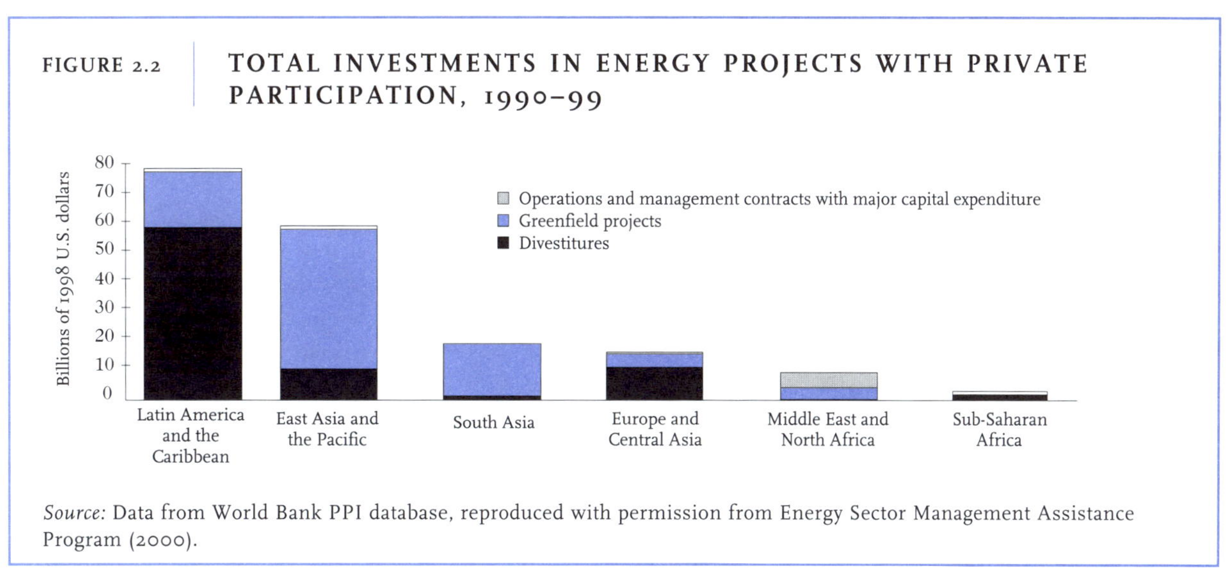

Source: Data from World Bank PPI database, reproduced with permission from Energy Sector Management Assistance Program (2000).

Institutional reforms across the developing world and transition economies were aimed at overcoming these hurdles, but took different forms in different parts of the world. *(See Figure 2.2.)* Private sector finance entered Latin America primarily through a wave of privatization. For countries burdened with debt-heavy utilities—the legacy of a wave of borrowing on international markets during the petro-dollar glut of the 1970s and 1980s—the outright sale of their public utilities was the most effective way to both shed debt and raise some capital (Oliviera, 1997).

In Asia, countries invited "independent power producers" (IPPs) to build and operate power plants and sell the electricity generated to the state utilities.[7] However, given the poor state of the sector in many countries, IPPs demanded and received highly favorable "power purchase agreements" (PPAs) providing concessions designed to minimize their risks and guarantee returns on their investment. These included high electricity prices; rigid "take-or-pay" contracts, which required the utility to make a set payment for power, whether it was used or not; and government guarantees against nonpayment for electricity. As a result, IPPs have been criticized as being built on the socialization of loss, but privatization of profit (Colley, 1997). Moreover, by

attracting capital only to generation, IPPs have potentially negative environmental implications, since they skew incentives toward new generation and against meeting electricity needs through greater efficiency. In addition, the purchase contracts have forced use of high-cost power over lower-cost power already available. Not least, suspicion over corrupt practices in striking these deals has tarnished the already dismal record of governance in the sector (Albouy and Bousba, 1998; Izaguirre, 1998).

In Africa, countries have embarked on electricity sector reform as part of a larger program of structural adjustment with a focus on public sector reform (Turkson, 2000). In addition to IPPs, the private sector often entered through management and operation contracts in which operation of the entire utility was handed over to a private entity. This approach was based on the small size of the sector in many African countries and the lack of strong regulatory frameworks (Covarrubias, 1996).

In Central and Eastern Europe, divestitures, along with some IPPs and a small number of management contracts, were the order of the day, as Figure 2.2 suggests. Divestitures were undertaken as part of a larger process of restructuring along the lines of the U.K. model. One important goal of electricity reform was to attract capital to replace and retool worn out systems. The reform process was further complicated by accession to the European Union and the consequent need to standardize systems and regulations, including environmental regulations (Chandler, 2000).

In sum, by the 1990s, development and transition economies sought to attract private capital to a sector shaped by decades of state ownership. In the past, state ownership was synonymous with running the sector to serve social policy—a "social compact." Indeed, many argue that good intentions led to an inefficient sector that has, in many countries, undermined the social purposes it sought to serve. With private capital as the taskmaster, however, the concern is a different one: How can a sector organized around private capital and markets sufficiently account for the public interest?

WHY PURSUE A PUBLIC BENEFITS AGENDA?

Will a reformed electricity sector organized around commercial principles automatically promote broader social and environmental interests? Or is there a case to be made for intentional scrutiny, oversight, and adjustment to ensure that these interests are adequately provided for? In this section we review this case on both social and environmental grounds. We conclude that, left entirely to its own devices, a market-oriented electricity sector provides inadequate long-run direction and runs the risk of providing insufficient public benefits.

Social Benefits: Access and Price

Electricity is a pervasive and central part of industrial society. Indeed, advocates for electricity sector reforms argue that the proposed changes will bring better quality power at lower average costs, with positive ripple effects through economies and societies. There may indeed be some benefits from efficiencies gained through application of the conventional reform model. However, when viewed through the lens of social equity, reforms should not only result in aggregate benefits, but also benefits to the least advantaged. From this perspective, access to electricity and the price at which electricity is available become important considerations. These issues are the central focus of this section.

Improved energy services provide a wide range of economic and social benefits, such as greater potential for education due to better lighting; savings in time and effort spent gathering traditional fuels; potential for improved access to information and digital connectivity; scope for greater productivity; scope for improved health services; and improved indoor air quality (Waddams Price, 2000; World Energy Assessment, 2000). In many developing countries, increasing access to electricity is an urgent need. In many African countries, for example, only 5 to 20 percent of the population has access to electricity with much of this access restricted to urban populations (Bhagavan, 1999).[8] In large measure, these dismal numbers reflect the failure of the centralized public power approach to guarantee access to electricity services.

There are several reasons for this failing. The rural poor, in particular, are often costly to serve because of remote locations and low population density, high transmission line losses, poor credit and minimal collateral, and a lack of purchasing and political power (Ehrhardt, 2000; World Energy Assessment, 2000). Yet, the proposed new model of market reform, designed to wring additional economic efficiency gains out of existing electricity networks through private sector competition, may be no better equipped to deal with these problems than the public utility model.[9] Indeed, in a reformed market where profitability is a central operating principle, interest in serving poor populations is likely to be further muted. A post-reform sector is likely to be dictated by principles of cost recovery in order to ensure adequate returns to the private sector. Hence, efforts at reform aimed at private participation will have to explicitly grapple with the tradeoffs between maintaining profitability and a social mandate to expand access to electricity, or risk being irrelevant to the problems of access.[10]

The limited evidence available to date supports the view that without an explicit effort, market reforms will not support greater access to electricity. In Bolivia, for example, a World Bank study concludes, "...the necessary expansion of the grid to connect the poor will not take place as a consequence of privatization and restructuring" (ESMAP, 2000). Indeed, the experience in industrialized countries suggests that rural electrification requires a distinct and directed effort. In the United States, for example, beginning in 1935 the Rural Electricity Administration ran a concerted program built around low-interest public funding, a model of rural electricity cooperatives, standardized engineering to reduce costs, and a principle of universal coverage that was highly successful in electrifying rural areas (McClean, 2000).

Because grid extension is technically and economically challenging, decentralized (usually small) electricity systems are increasingly becoming a feasible option for rural electrification. Among the available sources are several renewable energy technologies, including small hydropower, wind power, solar, and biomass (World Energy Assessment, 2000). Since distributed power sources avoid the high costs of transmission and distribution, these technologies may be cost-competitive for rural electrification. Leasing arrangements for technology, subsidies for the initial costs of switching to new systems, and concessionaire approaches (where a private entity is given exclusive right to a market in return for an obligation to serve) are all means of encouraging the spread of distributed electricity.

Whether through the grid or off-grid, therefore, there are various ways of promoting access even in an electricity sector led by private actors (Estache, Gomez-Lobo, and Leipziger, 2000). For example, at the time of privatization, distribution companies could be mandated connection targets, regulators could promote innovative approaches including concessionaire arrangements, and subsidies for the costs of connection to electricity services could help support the transition to electricity access. To meet the challenge of effectively serving dispersed populations, experiments with "micro-privatization," where delivery of electricity or other services is handed over to small scale-private or community actors, demonstrate better results than either large-scale public or private service delivery (Harper, 2000). On a case-by-case basis, the choice of appropriate mechanism or policy will depend on, for example, the existence of sufficient fiscal capacity, administrative capacity, and the scope for entry of competitive actors in each country (Ehrhardt, 2000). The general point is that mechanisms do exist to promote enhanced access to electricity. The contrasting examples of Morocco and Chile detailed in Box 2.5 suggest that this holds true for both public and private ownership, and further underscore the need for explicit mechanisms if enhanced access to electricity is to become a reality.

A related, concern is the price at which electricity will be available in a post-reform world. In industrialized countries, reforms will likely lead to lower prices as efficiency gains are captured. In developing countries, however, prices are likely to rise as current price restrictions are removed and cross-subsidies eliminated.[11] There is an important tension to be managed here. High prices are necessary to recover costs and to offer scope for profit in and attract private investment to the electricity sector. Yet, higher prices will disproportionately affect the poor, who tend to spend a larger proportion of their income on energy services than higher income groups.[12] Thus, public subsidies will likely continue to be necessary to meet social policy goals (Barnes and Halpern, 2000).

In this context, programs to encourage better end-use energy efficiency serve a valuable social purpose as well as an environmental one. Greater end-use efficiency would decrease the burden of price increases borne by households, and reduce the requirements for public subsidies to low-income consumers (Clark, 2000).

The salient point is that there are routes to promoting equitable social policies within a privatized electricity world. In order to win support for reforms, a sensible approach to promoting such policies would identify beneficiaries from existing implicit and explicit subsidies; examine the extent of access to electricity; use the information to design a program of mitigation; and inform the public of the program (Estache, Gomez-Lobo, and Leipziger, 2000). Largely

because of the political challenges, few countries follow this route.

There are two other related social concerns that are briefly treated here. First, the quality of electricity service in a post-reform sector is a significant component of a public benefits agenda. There is broad agreement that ensuring good service quality is central to the regulatory agenda in a post-reform sector, but there is no such consensus in the areas of access and price. Second, privatization is likely to be accompanied by substantial retrenchment of labor— an issue of great concern to public utilities unions (Colley, 1997; Bayliss and Hall, 2000). Moreover, since unions are typically well organized, labor

from households. The PERG program has increased the number of rural villages electrified annually from 557 in 1996 to 1,650 in 1999.

Chile's rural electrification program, launched in 1994, was designed to attract commercial participation by introducing competition at various levels. The central government, through the National Electricity Commission (CNE), transfers subsidy funds to regional governments. The central government allocates funds among regions based on progress in rural electrification the previous year, and the share of populations still lacking access to electricity. The selection of projects and companies to receive subsidies, however, is delegated to regional governments.

Regional governments pool their own budget resources with the subsidies allocated by the CNE. Private distribution companies, usually in partnership with municipalities or rural communities, submit project proposals to regional governments to obtain the bundled subsidies. Using criteria and tools developed by CNE, the responsible state agency identifies projects that qualify for subsidies (only

projects that generate a positive social return, but a negative private return are eligible), and submit these to the regional government council. Proposals are evaluated on the basis of cost-benefit analyses, share of the investment covered by the companies, and degree of social impact. From 1995 to 1999, the state contributed US$112 million in subsidies, but it also attracted US$60 million in private sector investments for rural electrification.

Sources:

Jadresic, Alejandro. 2000. "Auctioning Subsidies for Rural Electrification in Chile." Public Policy for the Private Sector. Note No. 214. Washington, D.C.: World Bank.

Jadresic, Alejandro. 2000. "A Case Study on Subsidizing Rural Electrification in Chile." In *Energy and Development Report 2000: Energy Services for the World's Poor*. Edited by Energy Sector Management Assistance Program. Washington, D.C.: World Bank.

Jamrani, Abderrahim. 1999. "The Moroccan General Electrification Programme (PERG)." Casablanca: Office National de l'Electricite (ONE).

Photovoltaic Market Transformation Initiative and the Global Environment Facility. 2000. *PVMTI News*. No. 2. September.

---

concerns offer a considerable stumbling block to the progress of reforms. Whether and how reforms address labor concerns is relevant both to the social dimensions of reform and to their political viability.

Environment: Balancing Efficiency and Sustainability

The electricity sector has a large environmental footprint. In much of the world, the sector is powered by fossil fuels, which after combustion emit local pollutants such as particulates and lead; regional pollutants such as sulfur dioxide; and global pollutants such as carbon dioxide. All of these

pollutants have impacts on human and ecosystem health (World Energy Assessment, 2000). While it is difficult to quantify the contribution of the electricity sector as distinct from all combustion of fossil fuels, Table 2.2 presents an estimate for one global pollutant—carbon dioxide (CO_2). Unfortunately, data for local and regional pollutants are unavailable. The data suggest that the electricity sector is a significant contributor to carbon emissions in all regions of the world.

How can reforms in the sector influence this relationship between electricity and environmental harms? From one perspective, reforms will automatically result in environmental improvements through

| TABLE 2.2 | **SHARE OF CARBON DIOXIDE EMISSIONS FROM ELECTRICITY AND HEAT PRODUCTION (1999)** |

Region	(%)
Middle East and North Africa	26
Sub-Saharan Africa	50
South Asia	50
Latin America and Caribbean	21
East Asia and Pacific	39
Europe and Central Asia	51
North America	42
High-Income Europe	30
Other High Income	40
World	**38**

Source: Computed by the WRI from International Energy Agency (IEA) data. *CO2 Emissions from Fossil Fuel Combustion.* Paris: OECD.
Note: Includes publicly or privately owned plants producing electricity or heat, for own use or for sale.

the full life-cycle costs of their generation technologies, and make decisions based on integration with their other energy assets as part of a strategic business plan (Sherry, 2000). The resulting decisions may be very different from, for example, an "Integrated Resource Planning" (IRP) approach, which seeks the least costly mix of options to meet energy service needs (Regulatory Assistance Project, 1994).[13] While IRP was originally designed for integrated utilities and may be challenging to implement in a market framework, the underlying point—the need to develop a long-run vision for the sector and suitable mechanisms for coordination—still holds. Planning mechanisms should, at a minimum, provide for the tradeoff between present and future benefits, include environmental sustainability, and make provisions for incentives in line with that vision. As concerns over both local and global environmental harms grow ever larger, reform in the electricity sector offers an opportunity to systematically promote a more sustainable energy future.

a number of pathways. The greater incentives for economic efficiency expected from a post-reform sector will translate into more efficient fuel use, with attendant environmental benefits. Incentives to shift to low-price fuels may stimulate a shift from coal-burning plants to relatively cleaner natural gas plants. And, since governments are typically more stringent about enforcing regulations with the private sector than the public sector, privatization will likely be accompanied by more rigorous enforcement of environmental regulation. These are all feasible outcomes, suggesting that reform may well result in a measure of environmental improvement.

However, there remain good reasons for close attention to the environmental implications of electricity market reforms. Market actors make decisions based on their business interests. While these may, on occasion, align with the interests of the sector as a whole, they are not guaranteed to do so. For example, private companies are likely to ignore

Indeed, in electricity sector reforms, some analysts foresee an opportunity to spur a transition to a "micropower" future—a sector based on small-scale generation units, of about the same size as the loads they power, and located closer to the end-user (Patterson, 1999; Dunn, 2000; Vaitheeswaran, 2001). Such a vision promises greater local control, reliability, and environmental benefits. Visions of a micropower future appear remote when juxtaposed against present realities in developing countries of inefficient and dirty centralized generation equipment, economically unviable pricing structures, and poor forms of governance. Yet, since the occasion to radically redesign institutions and promote innovation comes but seldom, it is important to seize such opportunities to put in place incentives for a more sustainable energy future.

While it is by no means clear what route a path from centralized power to micropower will take, it is likely that the process of electricity restructuring could put in place key preconditions for a micropower future based on renewable energy technologies. *(See Box 2.6.)* These include market access policies and price signals that reflect the true benefits of distributed power. Because innovative micropower systems will require new technological and institutional forms, the most promising opportunities to promote this approach may arise in currently underserved areas in the developing world, where the rigidities of the current system are not an obstacle (Patterson, 2001). In these areas, reforms in the electricity sector afford an opportunity to leapfrog to a new technological future.

Even if the goal is a somewhat more modest one of incremental improvements in environmental quality, an argument for separation of economic and environmental regulation, sometimes put forward by reform experts (Joskow, 1998), does not withstand scrutiny. Economic regulation can affect technology choice and can shape the transactions costs of different ways of supplying energy services. Economic regulators *de facto* make environmental decisions on a regular basis. Given this reality, is it not better that regulators are aware of and actively consider the environmental impact of their decisions?[14] At minimum, economic regulation should not discriminate against cleaner technologies.

<table>
<tr><td>BOX 2.6</td><td>**EXAMPLES OF POLICY OPTIONS TO DESIGN-IN ENVIRONMENTAL BENEFITS**</td></tr>
</table>

- *Environmental Dispatch:* Use a pollution index to "dispatch" power, in order to prioritize electricity from clean plants over dirty sources.

- *Standard Contracts:* Develop standardized contracts for power purchase to lower the negotiation costs to small renewable project developers.
  - Example: Small hydropower in Sri Lanka.

- *Price–driven Renewable Energy Incentives:*
  - Example: Avoided cost (U.S.) — utilities were required to purchase all the renewable power offered to them at the price it would otherwise have cost them to generate that power—the "avoided cost."
  - Example: Non-Fossil Fuel Obligation (U.K.) — a tax collected from fossil fuel generation was used to support renewable electricity plants.
  - Example: Feed-in law (Germany and Denmark) — sets a guaranteed price for the purchase of renewable energy.

- *Renewable Portfolio Standard:* A quantity-based policy that requires a specific amount of renewable energy to be purchased by all retail energy service providers.
  - Example: Several European countries and U.S. states.

- *Net Metering:* Use of a bi-directional meter to register electricity flow in both directions, making possible sale of extra power generated on site by distributed energy sources.

- *Public Benefits Fund:* Establish a fund created through a charge on transmission, for example, to be spent on public benefits such as wider access, energy efficiency, or development of sustainable energy technologies.
  - Example: PROCEL (Brazil) — concessionaires are required to spend 1 percent of revenues on energy conservation, and 0.25 percent on end-use efficiency.

Sources:

Hamrin, Jan. Forthcoming 2002. "Policy Options for 'Building-in' Environmental Protection at Different Stages of the Restructuring Process: Practical Advice on Clean Energy Policy." In *Energy Market Restructuring and the Environment*, edited by Martha Harriss. Lanham, MD: University Press of America.

World Energy Assessment. 2000. *World Energy Assessment: Energy and the Challenge of Sustainability*. Edited by United Nations Development Programme, United Nations Department of Economic and Social Affairs, and World Energy Council. New York: United Nations Development Programme.

For example, economic rules that allow open access to the transmission system are critical to encouraging independent suppliers to develop renewable energy sources. In India, such open "wheeling" policies have played a role in catalyzing a wind industry (Gupta, 2000).[15] Moreover, whether or not renewable energy competes on a level playing field may be determined by the details of electricity supply contracts, which specify issues such as how power can be purchased from remote locations and from intermittent sources such as renewable energy sources (Kozloff, 1998).

In another example, pricing of distribution services can either promote or hinder distributed generation (Regulatory Assistance Project, 1999). Since distribution costs can vary widely across a service area, power generated on site can bring considerable savings. Pricing of distribution services based on the average cost fails to send consumers appropriate signals on opportunities for cost savings. If regulators were to set the rates at which they buy power back from consumers to reflect the full cost of distribution, customers would have a greater incentive to invest in distributed power sources. These measures are particularly necessary since renewable energy technologies, which typically face high capital costs, may be at a disadvantage in raising capital under the conditions of volatile wholesale markets (Kozloff, 1998).

Efforts at energy efficiency may be a casualty of market reforms.

Finally, efforts at energy efficiency may well be a casualty of market reforms. For example, "unbundling" of utilities into generation, transmission, and distribution functions introduces transactions and information costs in the chain from production to consumption relative to a vertically integrated structure. These costs can erect severe barriers to promotion of end-use energy efficiency. For example, generating companies in an unbundled structure have no means of meeting new supply needs through end-use efficiency or demand management. For their part, distribution companies are unable to capture the full savings in transmission and generation created by greater efficiency, since these benefits are shared by other distribution companies. The resultant "free rider" problem is a disincentive to undertake efficiency enhancing programs (USAID, 1998). Evidence from restructuring efforts in industrialized countries suggests that these perverse incentives have not been addressed in reform programs. To make matters worse, industrialized country reforms have been accompanied by a decline in funding for energy efficiency programs (USAID, 1998).

This section has summarized arguments for a long-term vision for electricity, and the intertwined nature of economic regulation and social and environmental outcomes. The intent is not to argue for reforms to be driven by either social or environmental outcomes to the exclusion of economic concerns. However, taken collectively, these arguments present a strong case for not treating either social or environmental concerns as a residual, and for actively incorporating a public benefits agenda into the mainstream of reform efforts.

CONCLUSION

Electricity sector reform is emblematic of rapidly accelerating global integration. The nature of the debate has been transformed with the emergent dominance of market ideology, with institutional restructuring, and with accelerating flows of private capital into the sector. As with other debates about globalization, the future of the public interest depends on whether globalization will automatically take into account public concerns, or whether social and environmental interests will be explicitly factored into reform processes. This chapter has argued the latter, that market-led changes cannot automatically ensure the public interest. Having established that social and environmental concerns should be managed, the next question is how, and whether they are, in practice, internalized in reform processes. The following six chapters consider this question by drawing on the experience in a range of countries in the developing world and countries in transition.

NOTES

1. Small gas turbines generate electricity that costs about 4 cents per kilowatt-hour compared to 12 cents per kilowatt-hour for power from the nuclear plants completed in the late 1980s (Flavin and Lenssen, 1997). By one estimate, the minimum efficient plant size decreased from 1,000 megawatts in the early 1980s to between 50 megawatts and 350 megawatts by the late 1990s (International Energy Agency, 1999).

2. Motivations in both countries were similar: macroeconomic restructuring based on an ideological predisposition to private ownership and competition; a desire to increase efficiency in the sector; and privatization to stem a drain on public finances (Bacon, 1995; Rosenzweig and Voll, 1997; Patterson, 1999).

3. These changes were preceded by a number of measures introduced in the United States in the 1970s to address the changed context, including competition in generation, to boost new technologies, and to promote energy efficiency services aimed at the customer (Flavin and Lenssen, 1997; Patterson, 1999).

4. Indeed, those who initiated reforms in the United Kingdom admitted that even as they promoted competition, they had no clear idea of how competitive structures should be established (Hunt and Shuttleworth, 1996).

5. The following description of institutional reforms relies heavily on Hunt and Shuttleworth (1996).

6. In the wake of the Enron Corporation's problems, some analysts noted that deregulation was increasingly associated with high risk, and that some states in the United States were more actively considering retaining or returning to a central role in the sector (Johnson, 2002). Others argued that Enron's troubles are not tied to deregulation, and that an appropriate response would be to press forward with a deregulation approach based on greater competition and transparency (Vaitheeswaran, 2001).

7. Turkey was the first developing country to adopt this approach, which was a substantial departure from the then prevalent approach of building a power plant and handing it over to the government for operation (Patterson, 1999).

8. Specifically, in Malawi, 4 percent of the population has access to electricity; in Tanzania, 6 percent; in Uganda, 12 percent; and in Zimbabwe, 14 percent (Bhagavan, 1999).

9. As Brook and Besant-Jones (2000) put it, "… it is arguable that the poorest of the poor, who make up the majority of the estimated 2 billion people who do not have access to modern energy, do not stand to benefit much from reforms aimed primarily at existing electricity and gas networks."

10. Privatization could also lead to a decrease in access to electricity unless potential problems are anticipated up front and mitigated. In Argentina, privatized distribution companies shut off supply to the poorest urban neighborhoods and emergency settlements in order to reduce losses (Bouille and Dubrovsky, 2000). Similar experience with disconnection of supply following privatization has been reported from Georgia, Moldova, and the Dominican Republic (Bayliss, 2001).

11. In the longer term, efficiency improvements may hold down costs and lower prices.

12. Moreover, as happened in Argentina, private sector firms will have an incentive to discourage consumption by their poor and less reliable consumers by increasing their prices, and decreasing those of the wealthy (Bouille and Dubrovsky, 2000).

13. According to one view, IRP is a necessary complement to competition because price only provides information on what a resource costs, not on its full worth to a utility (Reddy and Sumithra, 1997). By focusing on the "avoided cost" of supply, IRP provides a framework to compare the true worth of energy sources from both the demand and supply sides. For example, a source of power that has a higher price than another may also have a worth that justifies the price premium if it provides largely peak power, or if it saves on transmission and distribution costs.

14. As the Regulatory Assistance Project—a body of former U.S. government officials with considerable experience in regulatory theory and practice—compellingly argues, no utility regulator would urge their environmental counterpart to remain ignorant of the economic effects of environmental regulation (Regulatory Assistance Project, 1999).

15. Moreover, whether or not renewable energy competes on a level playing field may be determined by the details of electricity supply contracts, which specify issues such as how power can be purchased from remote locations and from intermittent sources such as renewable energy sources (Kozloff, 1998).

REFERENCES

Albouy, Yves, and Reda Bousba. 1998. "The Impact of IPPs in Developing Countries—Out of the Crisis and into the Future." Viewpoint Note No. 162 (December). Washington, D.C.: World Bank.

Asian Development Bank. 1994. "Bank Policy Initiatives for the Energy Sector." Manila, Philippines.

Bacon, R.W. 1995. "Privatization and Reform in the Global Electricity Supply Industry." *Annual Review of Energy and Environment.* 20:119-143.

Bacon, Robert. 1999. *Global Energy Sector Reform in Developing Countries: A Scorecard.* Report No. 219-99. Washington, D.C.: UNDP/World Bank.

Bacon, Robert, and John Besant-Jones. 2001. "Global Electric Power Reform: Privatization and Liberalization of the Electric Power Industry in Developing Countries." *Annual Review of Energy and Environment* 26:331-359.

Balu, V. 1997. "Issues and Challenges Concerning Privatisation and Regulation in the Power Sector." *Energy for Sustainable Development* III. (6):6-13.

Barnes, Douglas F., and Jonathan Halpern. 2000. "The Role of Energy Subsidies." In *Energy Services for the World's Poor,* edited by Energy Sector Management Assistance Program, Washington, D.C.: World Bank.

Bayliss, Kate. 2001. "Privatisation of Electricity Distribution: Some Economic, Social and Political Perspectives." London: Public Services International Research Unit. April. Online at: http://www.psiru.org/reports/2001-04-E-Distrib.doc. (May 20, 2002).

Bayliss, Kate and David Hall. 2000. "Privatisation of Water and Energy in Africa." London: Public Services International Research Unit. September. Online at: http://www.psiru.org/reports/2000-09-U-Afr.doc. (May 20, 2002).

Besant-Jones, John. 1996. "The England and Wales Electricity Model—Option or Warning for Developing Countries." Viewpoint Note. No. 84. Washington, D.C.: World Bank.

Bhagavan, M.R. 1999. "Introduction." In *Reforming the Power Sector in Africa,* edited by M. Bhagavan. London: Zed Books.

Bouille, Daniel and Hilda Dubrovsky. 2000. "Reform of the Electric Power Sector in Developing Countries: Case Study of Argentina." Unpublished manuscript. Buenos Aires: Institute of Energy Economics, Fundación Bariloche.

Brook, Penelope J. and John Besant-Jones. 2000. "Reaching the Poor in an Age of Energy Reform." In *Energy Services for the World's Poor,* edited by Energy Sector Management Assistance Program. Washington, D.C.: ESMAP.

Chandler, William. 2000. *Energy and Environment in the Transition Economies.* Boulder, CO: Westview Press.

Clark, Alix. 2000. "Making Provision for Energy Efficiency Investment in Changing Electricity Markets: International Perspectives." Paper read at IEI Workshop on Public Benefits and Power Sector Restructuring, Cape Town, South Africa, April 10.

Colley, Peter. 1997. *Reforming Energy: Sustainable Futures and Global Labour.* London: Pluto Press.

Covarrubias, Alvaro J. 1996. *Lending for Electric Power in Sub-Saharan Africa.* Washington, D.C.: World Bank.

Dunn, Seth. 2000. *Micropower: The Next Electrical Era.* Washington, D.C.: Worldwatch Institute.

Ehrhardt, David. 2000. "Impact of Market Structure on Service Options for the Poor." Paper read at conference, Infrastructure for Development: Private Solutions and the Poor, London, U.K., May 31 - June 2.

Energy Sector Management Assistance Program. 2000. *Introducing Competition into the Electricity Supply Industry in Developing Countries: Lessons from Bolivia.* Report 233/00. Washington, D.C.: ESMAP.

Estache, Antonio, Andrews Gomez-Lobo, and Danny Leipziger. 2000. "Utilities 'Privatization' and the Poor's Needs in Latin America: Have We Learned Enough to Get it Right?" Paper read at conference, Infrastructure for Development: Private Solutions and the Poor, London, U.K., May 31 - June 2.

Flavin, Christopher and Nicholas Lenssen. 1997. "Powering the Future: Blueprint for a Sustainable Electricity Industry." *Quarterly Bulletin.* 18 (3).

Graham, Stephen and Simon Marvin. 1995. "More than Ducts and Wires: Post-Fordism, Cities and Utility Networks." In *Managing Cities: The New Urban Context,* edited by P. Heal, S. Cameron, S. Davoudi, S. Graham and A. Madani-Pour. New York: John Wiley & Sons.

Gupta, Ajit K. 2000. "Policy Approaches: The Indian Experience." Paper read at conference, Accelerating Grid Based Renewable Energy Power Generation for a Clean Environment, Washington, D.C., March 7-8.

Harper, Malcolm. 2000. *Public Services through Private Enterprise*. London: Intermediate Technology Publications.

Hirsh, Richard. 1999. *Power Loss: The Origins of Deregulation and Restructuring in the American Electric Utilities System*. Cambridge, MA: MIT Press.

Hunt, Sally, and Graham Shuttleworth. 1996. *Competition and Choice in Electricity*. New York: John Wiley and Sons.

International Energy Agency. 1999. *Electricity Market Reform: An IEA Handbook*. Paris: OECD/IEA.

Izaguirre, Ada Karina. 1998. "Private Participation in the Electricity Sector—Recent Trends." Viewpoint Note No. 154. Washington, D.C.: World Bank.

Johnson, Kirk. 2002. "Turn out the Lights, the Party's Over: After Enron, Deregulation Looks More Risky than Sexy." *New York Times*. (February 10):29.

Joskow, Paul L. 1998. "Electricity Sectors in Transition." *Energy Journal*. 19 (2):25-52.

Kozloff, Keith. 1998. "Electricity Sector Reform In Developing Countries: Implications for Renewable Energy." REPP Research Report No. 2. Washington, D.C.: Renewable Energy Power Project.

Littlechild, Stephen. 2000. *Privatization, Competition and Regulation in the British Electricity Industry, with Implications for Developing Countries*. Report 226/00. Washington, D.C.: World Bank.

Magnus, Eivund. 1997. "Competition without Privatization—Norway's Reforms in the Power Sector." *Energy for Sustainable Development* III. (6):27-35.

McClean, Christopher. 2000. Paper read at conference, Village Power 2000, Washington, D.C., December 4-7.

Newbery, David. 1995. "A Template for Power Reform." Viewpoint Note 54. Washington, D.C.: World Bank.

Newbery, David M. and Richard Green. 1996. "Regulation, Private Ownership and Privatisation of the English Electricity Industry." In *International Comparisons of Electricity Regulation*, edited by R. Gilbert and E. P. Kahn. Cambridge: Cambridge University Press.

Oliviera, Adilson de. 1997. "Electricity System Reform: World Bank Approach and Latin American Reality." *Energy for Sustainable Development* III. (6):27-35.

Oliviera, Adilson de and Gordon MacKerron. 1992. "Is the World Bank Approach to Structural Reform Supported by Experience of Electricity Privatization in the UK?" *Energy Policy* (February):153-162.

Patterson, Walt. 1999. *Transforming Electricity*. London: Earthscan.

———. 2001. "Introduction." Paper read at conference, Keeping the Lights On: Electric Tradition or Innovation?, London, U.K., March 14-15.

Reddy, Amulya K N. 2001. "California Energy Crisis and its Lessons for Power Sector Reform in India." *Economic and Political Weekly* XXXVI (May 5-11).

Reddy, Amulya K.N. and Gladys Sumithra. 1997. "Integrated Resource Planning." *Energy for Sustainable Development* III (6):14-16.

Regulatory Assistance Project. 2001. *IRP and Competition*. Regulatory Assistance Project 1994. Online at: http://www.rapmaine.org/irp.html. (April 21, 2001).

———. 2001. *Electric Industry Restructuring and the Environment*. Regulatory Assistance Project 1999. Online at: http://www.rapmaine.org/environment.html. (April 21, 2001).

Rosenzweig, Michael B. and Sarah P. Voll. 1997. "Sequencing Power Sector Privatization." Unpublished manuscript. Washington, D.C.: National Economic Research Associates.

Sherry, Christopher John. 2000. "Economic Globalization, the Environment, and the Electric Power Industry: U.S. Foreign Direct Investment in the Greenhouse." Masters thesis, Center for Energy and Environmental Policy, University of Delaware, Newark, DE.

Tellam, Ian, ed. 2000. *Fuel for Change: The World Bank Energy Policy, Rhetoric vs. Reality*. London: Zed Books.

Teplitz-Sembitzky. 1990. *Regulation, Deregulation or Reregulation: What is Needed in the LDCs Power Sector Industry and Energy Departments?* Washington, D.C.: World Bank.

Turkson, John K., ed. 2000. *Power Sector Reform in Sub Saharan Africa*. London: MacMillan Press.

United States Agency for International Development. 1998. *Case Studies of the Effects of Power Sector Reform on Energy Efficiency*. Report No. 98-03. Washington, D.C.: USAID.

———. 1998. *Promoting Energy Efficiency in Reforming Electricity Markets*. Report No. 98-02. Washington, D.C.: USAID.

Vaitheeswaran, Vijay. 2001. "A Brighter Future?" *Economist*. February 10.

———. 2001. "Electricity Deregulation is Still Sound Policy." *New York Times*. December 15.

Waddams Price, Catherine. 2000. "Better Energy Services, Better Energy Sectors—and Links with the Poor." In *Energy Services for the World's Poor*, edited by Energy Sector Management Assistance Program, Washington, D.C.: World Bank.

Watts, Price C. "Heresy? The Case Against Deregulation of Electricity Generation." *The Electricity Journal*. 2001 (May):19-24.

Wolak, Frank A. No Date. *Market Design and Price Behavior in Restructured Electricity Markets: An International Comparison*. Department of Economics, Stanford University. On line at: http://www.stanford.edu/~wolak (March 22, 2002).

World Bank. 1988. "A Review of World Bank Lending for Electric Power." Energy Series No. 2. Washington, D.C.: World Bank.

———. 1993. *The World Bank's Role in the Electric Power Sector*. Washington, D.C.: World Bank.

———. 2001. "The California Experience with Power Sector Reform." Unpublished manuscript. Washington, D.C.: World Bank. Online at: http://www.worldbank.org/html/fpd/energy/pdfs/e_calexpo400.pdf (May 29, 2002).

World Energy Assessment. 2000. *World Energy Assessment: Energy and the Challenge of Sustainability*. Edited by United Nations Development Programme, United Nations Department of Economic and Social Affairs, and World Energy Council. New York: United Nations Development Programme.

3

ARGENTINA

MARKET-DRIVEN REFORM OF THE ELECTRICITY SECTOR

Daniel Bouille
Hilda Dubrovsky
Crescencia Maurer

INTRODUCTION

In 2001, Argentina garnered considerable international attention as the country's government took increasingly desperate measures to jump-start its economy. Mired in four years of recession, Argentina sought to avert the devaluation of the Argentine peso, and prevent outright default on billions in debt obligations to international banks and bond-holders. The current economic crisis is reminiscent of another that gripped Argentina in the late 1980s and also precipitated the downfall of a president. These two crises frame Argentina's experience with electricity sector reform.

In the 12-year period between crises, Argentina radically changed how the electricity sector was organized and the public sector's role in the supply of electricity services. In the early 1990s, new regulated and competitive markets were created to allow private actors to enter and supply electricity services previously owned and operated by federal and provincial utilities. The federal government, followed eventually by a number of provincial governments, redefined its role. It set the framework for markets to operate and regulated their operation in the public and private interest. In the context of this report, Argentina is of particular interest both because of its wholesale adoption of market reforms and the length of its experience with their implementation. For a profile of the electricity sector in Argentina see Box 3.1.

Electric power reforms in Argentina created a significant number of public benefits, among them

<table>
<tr><td>BOX 3.1</td><td>PROFILE OF THE ELECTRICITY SECTOR IN ARGENTINA</td></tr>
</table>

Population (2001)[1]: 37 million

Population with access to electricity (2000)[2]:
Total: 95%   *Rural:* 70%   *Urban:* 98%

Installed electricity generation capacity (1999)[3]
Total: 23 gigawatts (0.72% of total world capacity)
     Thermal: 57%
     Hydro: 39%
     Nuclear: 4%
     Geothermal and Other: 0%

CO_2 emissions from electricity and heat as a share of national emissions (1999)[4] : 22%

Notes:

1. World Resources Institute. 2000. *People and Ecosystems: The Fraying Web of Life.* Washington, D.C.: World Resources Institute.

2. http://eclac.org/publicaciones/Poblacion/2/LCG2052/BD6311.html (June 17, 2002).

3. Argentine Energy Secretary.

4. Computed by WRI using International Energy Agency (IEA) data. IEA. 2001. *CO2 Emissions from Fossil Fuel Combustion.* Paris: OECD.

improved quality of service in urban areas, reductions in technical and non-technical losses, expansion of supply in urban areas and increased efficiency. Nevertheless, other public benefits failed to materialize. Evidence points to an absence of incentives to increase energy efficiency among distribution companies and household consumers, limited expansion of electricity access to the most isolated rural populations, and a failure to correct regulatory obstacles that prevented expansion of the transmission system.

Despite Argentina's relative success with the introduction of market reforms in the electricity sector, policymakers and donors recognized (in the former case only reluctantly) that a number of public benefits were underserved by the newly reformed electricity markets (World Bank, 2000). Second generation reform efforts to address problems with public benefits consisted largely of tinkering with or refining the operation of the competitive and regulated markets. There was little political interest in revisiting the reforms themselves to better protect public benefits.

All reform processes such as those implemented in Argentina are dynamic in nature and respond to changing circumstances and challenges. This chapter assesses whether reformers responded to sustainable development concerns when they established the direction of reform, in its subsequent implementation, and in follow-on efforts to introduce additional reform. In other words, how and when were social and environmental issues raised during the 12-year period?

BACKGROUND

The privatization and reform of Argentina's electric power sector in the early 1990s reflected a radical change in vision about the role of the public sector in economic development. In 1989, hyperinflation, crushing debt, and poor public services led to the resignation of President Raul Alfonsín five months before the end of his term. Carlos Saul Menem, already the president-elect, stepped into office. The new president entered into a political agreement with the opposition that facilitated congressional passage of two laws that were key precursors to the electricity reforms. These laws were the State Reform Act, which gave the executive extraordinary powers to reorganize

and privatize public enterprises, and the Economic Emergency Act, which suspended subsidies and lifted barriers to foreign investment (Abdala, 2001). With their passage, Menem embarked on a wholesale reform program that had as its centerpiece the introduction of competitive markets and the reduction of a direct public sector role in key economic sectors, including electricity (World Bank, 1995).

Policymakers and donors recognized that a number of public benefits were underserved by the newly reformed electricity markets.

For the Menem administration, the electricity sector illustrated the inefficiencies and problems with the public sector's monopoly provision of economic services. In the late 1980s, federal utilities experienced successive brownouts, fluctuating voltage levels, and electricity shortages. These problems were worsened by a drought that restricted hydropower generation—one of the country's main sources of electric power. In addition, federal utilities constituted a significant drain on the federal budget; their operating costs and debts far exceeded the revenues they generated. Repeated efforts to "corporatize" these utilities had proven unsuccessful. Vested interests, including the utilities' own technicians and bureaucrats, trade unions, federal and provincial politicians, and private suppliers and contractors limited the effect of such efforts.[1]

The reformers selected by Menem—principally Domingo Cavallo, the Economy Minister—believed that the solution to the electricity sector's problems was the introduction of competitive markets. This would be achieved mainly through vertical and horizontal unbundling of the power chain—generation, transmission, and distribution—and their subsequent privatization. They believed these changes would bring market discipline to the sector and eliminate the drain on the federal budget.

Over a two-year period (1990-1991), a small team in the Secretary of Energy carried out technical

studies and developed rules and operational guidelines that formed the basis for the electricity sector's
restructuring. In 1992-93, the reform process
accelerated with the passage of the Electricity
Regulation Act, the privatization of federal utilities,
and the creation of a new sectoral regulatory body
known as the National Entity for Electricity Regulation (ENRE). The speed of the reform was such that
generation and distribution assets were privatized
before ENRE began to operate (Abdala, 2001). Box
3.2 provides a chronology of major legal and
regulatory changes in the sector.

The new structure drew from experiences with
power sector reform in the United Kingdom (vertical
and horizontal unbundling of the sector), and Chile
(an open access wholesale market, marginal cost
pricing of wholesale electricity, and deregulation of
large power consumers) (Besant-Jones, 1996).
Argentina introduced innovations of its own, such as
limiting concentration of ownership both across
segments (vertical) and within segments (horizontal),
and the introduction of a sector-specific regulator
(Estache and Rodriguez-Pardina, 1999). The structure of the post-reform electricity sector is illustrated
in Figure 3.1.

Under the reformed structure, government retains a
policymaking role and participates in the operation of
the wholesale electricity market managed by an
independent nonprofit corporation, CAMMESA,
jointly owned by the federal government (represented
by the Secretary of Energy), power generators, trans-
mitters, distributors, large consumers, and brokers.
The mandate of the sector regulator, ENRE, included
establishing safety and operating standards, setting
tariffs for transmission and distribution, enforcing
standards and laws that apply to distribution and
transmission companies, resolving disputes in the
sector, and conducting public hearings on regulatory
reforms. Five commissioners govern ENRE. Each is
selected through a competitive nomination process,
and is subject to approval by Congress.

The reforms allowed the regulated entry of private
suppliers into the transmission market. The federal
government auctioned transmission concessions for
the existing 500-kilovolt lines, setting a price cap
concession holders can capture from charges on
generators selling electricity in the wholesale market
(Estache and Rodriguez-Pardina, 1999). New
transmission lines, however, were to be financed in a
different manner. In such cases ENRE approves a
build, operate, and maintain (BOM) contract that
allows costs to be recouped through charges on
beneficiaries. Alternatively, ENRE can approve a
private contract between a generator and the consumers who will use the added transmission capacity. Under the first option, beneficiaries are defined
as the network of users located on a node where
electricity flows will change as a result of the new
project. If beneficiaries representing at least 30
percent of the pool contest the proposed transmission charge, ENRE blocks the proposed project
(Abdala and Chambouleyron, 1999).

The distribution assets of the federal electricity
sector were divided into three 99-year concessions—
2 in Buenos Aires and 1 in La Plata—with 10-year
intervals for the revision of price caps, and the option
for concession holders to give up the concession and
allow a new competitive bidding process. Distribution concession contracts include requirements for
universal service to all residential and small consumers, but also guarantee a concessionaires' monopoly.

After reforms and privatization were complete at
the federal level, the Secretary of Energy, led by
Carlos Bastos, sought to extend reforms to the
provinces. Under Argentina's federal system,
provincial authorities retain regulatory and
policymaking powers and can structure ownership of
local generation and distribution assets as they see
fit. Beginning in 1993-94, the federal government
used the power of the purse to push provinces to
follow the federal reforms. The transfer of federal
funds—including provinces' shares of the National
Electricity Fund, federal fuel taxes, and end user
tariffs—was conditioned on provincial governments'
conformance with the federal pricing structure
(Pistonesi, 2000). Only those provinces able to
operate their utilities without these transfers retained
a measure of autonomy. By 2001, 14 of 24 provinces
had privatized their distribution assets.

CHRONOLOGY OF ELECTRICITY SECTOR REFORM IN ARGENTINA

1987 — The Secretary of Energy and Economy passes Resolution No. 475 calling for the development of environmental norms and regulations for the energy sector.

1987 — Government issues official environmental management handbook for hydroelectric projects.

1988 — Government issues official environmental management handbook for high-tension transmission lines.

1988–89 — Electricity supply crisis.

1989 — New presidency assumes executive branch (Menem administration).

1989 — Administrative Reform Law No. 23696 establishes the basis for privatization of all state-owned companies.

1990 — Government issues official environmental management handbook for conventional central thermoelectric generating plants.

1991 — The World Bank grants the Government of Argentina a $300 million loan to assist with public sector reforms and privatization of state companies in the telecommunications, railroad, and fossil fuel sectors. This loan included funds to assist with privatization in other sectors.

1991 — Decree No. 634 issued on the reconversion of the electric power sector. This decree establishes a wholesale market, defines final consumers, and unbundles generation, transmission, and distribution functions.

1992 — Law No. 24065, the Electricity Regulation Act, comes into force and assigns normative responsibilities to the Secretary of Energy. These responsibilities include environmental enforcement, application of environmental management handbooks, and establishing emission limits for thermal generating plants.

1992 — Resolution No. 61, Organization of the Electric System, defines private agents and procedures for the function of the electricity market.

1992 — The National Entity for Electricity Regulation (ENRE) is created. Environmental regulation is assigned to public security entities required to enforce specific regulations and apply penalties (articles 77 and 78).

THE REFORM PROCESS IN ARGENTINA

The conceptualization and implementation of the initial reform program was carried out by a small group of policymakers and financial, legal, and technical consultants in the Secretary of Energy, which was part of the Ministry of Economy and Public Works. This team, including the Energy Secretary, Carlos Bastos, shared the Minister's disciplinary and technical background, and his commitment to market-based reforms. The group's homogeneity, the urgency of the national economic crisis, and the closed nature of the reform process did not allow for internal or external debate about the speed or content of the reform.

During this period, few political or economic forces could challenge the Menem administration's premises or plans. The Menem administration worked

(CONTINUED)

- **1992–93** — Federally owned power plants and distribution companies are privatized. Distribution concessions eliminate subsidies.

- **1993–97** — Concessions awarded to companies to construct hydroelectric plants and expand the transmission system.

- **1993–2002** — 14 of 24 provinces privatize distribution companies in line with federal reform.

- **1994** — Resolution No. 6, the "Four-Year Framework Agreement," is adopted. It establishes a four-year time period for reductions of illegal or irregular consumption. As a result, approximately 650,000 consumers are formally connected to the grid system.

- **1994** — Resolution 159/94, issued by the Secretary of Energy, creates a regime for large consumers (those consuming between 100 kilowatts and 2 megawatts) allowing them to purchase electricity directly from generators.

- **1995–98** — The PAEPRA Program to supply electricity to isolated rural areas is designed and receives the support of the Global Environment Facility (GEF) and the World Bank.

- **1998** — A resolution is issued by the Secretary of Energy that reduces the floor of what constitutes a large consumer (to 50 kilowatts) and allows them to establish defined supply contracts with a generator.

- **1999** — A large electricity blackout in the EDESUR distribution concession area affects more than 160,000 in the federal capital, in some cases for over 10 days.

- **2000** — Transmission connections are established with Brazil and agreements are finalized to permit the export of 1,000 megawatts.

- **2000** — De la Rua administration takes office and initially endorses a federal role in the expansion of the national electric transmission system proposed by Energy Secretary Mac Karthy in the last Menem administration.

- **2001** — Economy Minister Cavallo assumes reins of Argentine economy and rescinds order that established a federal role in funding expansion of the transmission system.

- **2001** — Argentine economic crisis leads to the resignation of President de la Rua and Economy Minister Cavallo.

with congressional members from its own Peronist party to form a legislative alliance that drafted and passed the necessary framework legislation for the electric sector reforms. The legislative branch neither opposed nor critically reviewed the executive branch's proposals. Furthermore, through the passage of the State Reform Act, the Congress effectively limited its oversight role to that of providing broad direction rather than contributing substantive decisions about how to privatize or regulate industries or services (Abdala, 2001).

It should not be assumed, however, that there was complete agreement with the federal government's diagnosis of the sector's problems or the proposed market solutions.[2] Academic and research communities and some donors maintained that the sector had operated efficiently in the hands of the state for many years. They pointed out that state control achieved self-sufficiency, expanded modern energy sources, established an adequate balance between reserves and consumption, and guaranteed future supplies of electric power. This group believed the sector's

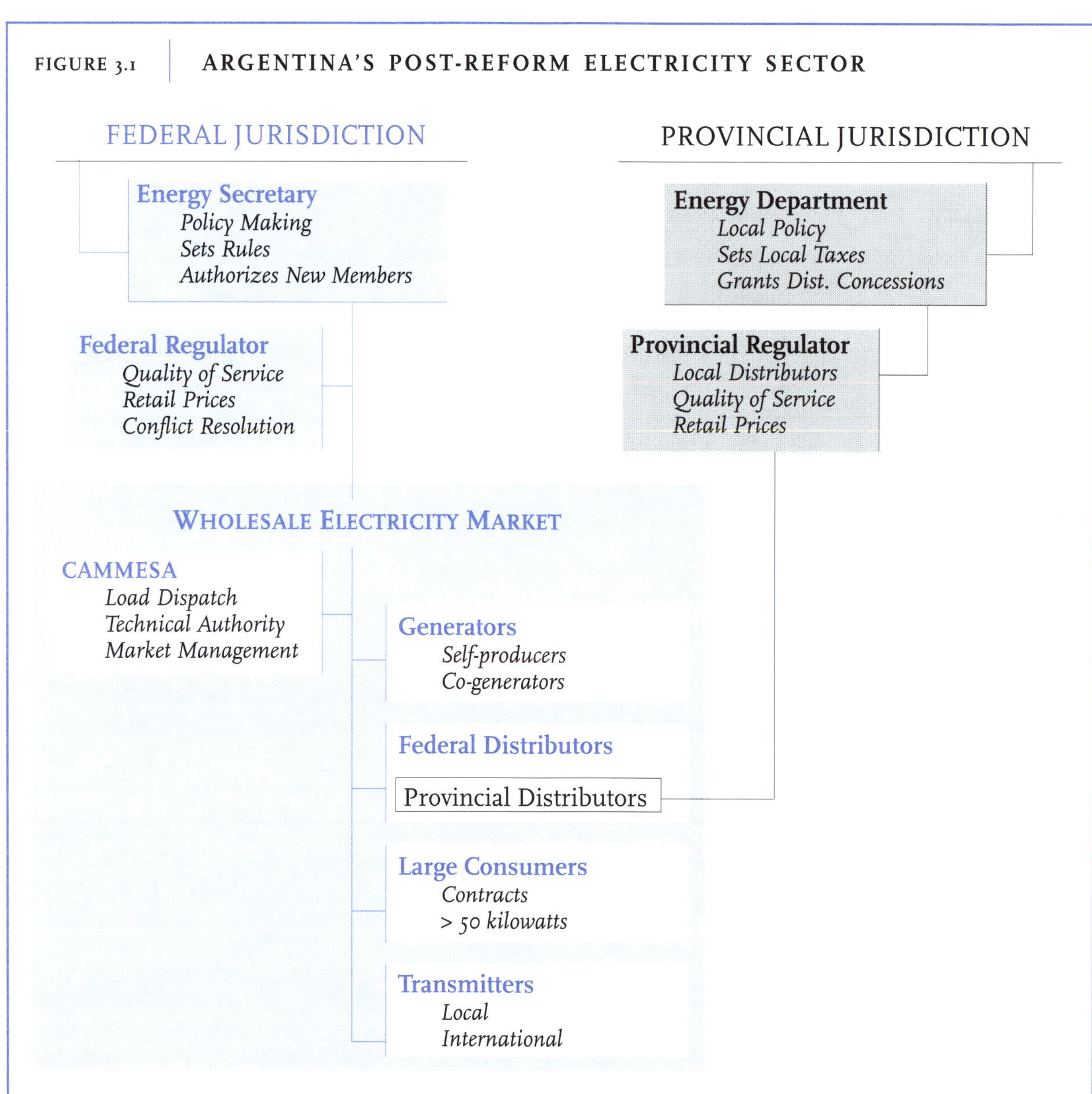

problems were caused by developments that were external to the sector—including high levels of foreign debt, economic recession, and drought—but which affected its fiscal and technical performance. Given the speed and closed nature of the reform process, these viewpoints did not influence the decisions taken by the Bastos team (World Bank, 2000).

Among civil society organizations, Argentina's labor unions had a long history of activism and held considerable political power. Over 120 state-owned enterprises were privatized in the early 1990s; surprisingly, only some unions opposed the privatization. One of the reasons was that employees of privatized enterprises were granted 10 percent of the new firms' equity shares, allowing labor groups

to capture economic rents from the process
(Margheritis, 1999). More importantly, the majority
of unions were closely associated with the Peronist
Party, Menem's political party. Consequently, unions
supported the reforms to preserve their historic
political alliance and to capture rents.

The Secretary of Energy also marketed Argentina's
privatization program to international banks, private
electric utilities, and independent power producers
in the United States and Europe (Bastos and Abdala,
1996). The government employed consultants,
including a number from Wall Street firms, to
produce financial evaluations of the new business
units to be created from existing federal utilities, and
to explain the newly restructured sector's policies and
regulations. The government's principal concern was
to demonstrate that it was re-establishing macroeconomic stability, committed to a market-based approach, and deserving of better financing terms from
capital markets (Bastos and Abdala, 1996).

In addition to this marketing effort, the government sold the federal utilities on terms that were
quite favorable for private buyers. It set no floor on
the price bidders could offer for these utilities. It
accepted debt securities at below-market values as
payments for shares in newly privatized companies,
and released buyers from obligations to honor debt
guarantees that utilities had taken on before
privatization. The overriding concern of the reformers was to attract and retain private investment in the
newly privatized utilities (Hasson, 1994).

The government frequently asserted that the
reforms would dramatically increase economic
efficiency, which in turn would generate positive

social and environmental spillover effects. Private
provision of electricity would increase competition
and efficiency, and lead to higher quality but lower
cost electricity services. The government also argued
that the reforms would permit it to focus on its
regulatory and enforcement roles (Bastos and
Abdala, 1996). Not surprisingly, the Bastos team did
not formally consult with other branches of government or civil society groups on the reform plans; the
need to resolve the country's economic crisis justified
proceeding to immediate implementation.[3]

The Role of the Multilateral Development Banks

At the initial stages of reform (1990-93), the stron-
gest donor support for Argentina's power sector
reforms came from the World Bank. During the
1970s and 1980s, the World Bank had provided loans
for government ownership and development of
integrated utilities. By the early 1990s, that institution was reassessing the performance of publicly
dominated electricity sectors and coming to the
conclusion that market competition should be
introduced to increase their efficiency, cover operating costs, and attract the necessary investments to
fund expansion of generating capacity (World Bank
and OLADE, 1991). (*See Box 3.3.*)

To speed resources to the government, the World
Bank amended loan agreements originally intended
to support improvements in the operational and
managerial performance of the three main federal
utilities (SEGBA, Hidronor, and Agua y Electricidad).
These funds were reprogrammed to permit payment
of consultants and support staff within the Secretary
of Energy who were developing the reform and
privatization plan. Because this involved an amendment rather than approval of an existing loan
agreement, no conditionalities were placed on
expenditures or disbursements.[4]

The Inter-American Development Bank (IDB) also
provided considerable support for the power sector
reforms, but it did so around 1994, a few years after
the World Bank, when the reform and privatization

program was already defined and the process of implementation under way. IDB staff were more critical of the Argentine power sector reforms than were World Bank staff. Some sector specialists did not agree with the privatization of federal utilities. And, in general, IDB staff believed that sector reforms should be completed before the privatization of federal utilities.[5] Although the IDB provided a $300 million fast-disbursing sector loan to assist the Secretary of Energy, this loan was directed at reform efforts rather than privatization. Conditionalities placed on this loan included demonstrated progress in the definition of the Secretary of Energy's policymaking role after privatization, the operation of ENRE, and development of the Secretary of Energy's environmental and social assessment unit.[6]

Provinces were reluctant to emulate federal reforms because they garnered considerable rents from local utilities and they were convenient vehicles for political patronage.

At early stages in the process, both the World Bank and the IDB believed that significant public benefits would flow from both sector reforms and privatization. Staff at both banks believed that improving the economic performance and efficiency of the federal utilities would generate positive externalities such as improvements in local air quality, higher quality electricity services at lower wholesale and retail prices, and increased private investment in new generation capacity.[7]

In the years immediately following privatization of the federal utilities (1993-94), the Bastos team worked in concert with the World Bank and the IDB to push provincial governments to adopt the federal model of sector reform and privatization. Many provinces were reluctant to emulate federal reforms because they garnered considerable rents from local utilities and they were convenient vehicles for political patronage.

The assistance the IDB and the World Bank offered to provincial governments included conditions imposed by the federal government. The federal government would not approve multilateral development assistance to the provinces unless they agreed to conform to the federal reform scheme, in particular the tariff and electricity pricing structure established for distribution concessions (Pistonesi, 2000). Other conditions imposed by the federal government or the banks included creation of independent provincial regulators, reform of electricity supply structures, and reduction of employees in provincial enterprises.[8]

The provinces that were most dependent on federal or donor funds generally adopted the federal model. Other provincial governments followed the federal model, but made significant adjustments such as requiring holders of distribution concessions to make mandatory investments or refusing to index electricity prices to U.S. inflation. The conditions imposed by the banks and the federal government proved only partially successful. To date, 10 of 24 provinces—among them Cordoba and Santa Fe, which are large and account for significant electricity consumption—have not begun, or have interrupted the privatization of their distribution and transmission services.

Interestingly enough, the measures the federal government and the World Bank recommended to provincial governments implicitly recognized a number of shortfalls in the federal reform process. Provinces were encouraged to take the following actions:

- Maximize the benefits of market competition;
- Offer new owners incentives to take actions consistent with the public interest;
- Include contract stipulations that state as clearly as possible all rights and obligations;
- Allow room for the renegotiation of contracts;
- Spell out the responsibilities of private owners when transferring public assets;

- Avoid underestimating the importance of tariff design; and
- Give equal attention to provincial regulatory capacity and to concession licenses and contracts (World Bank, 1996).

These recommendations actually represented stricter or more rigid guidelines for privatization than those implemented at the federal level.

PUBLIC BENEFITS

In Argentina, the public benefits that generated the greatest controversy and public attention in the 1990s were those generally defined as "social benefits." The most salient of these included access to and quality of service, the distribution of the costs and benefits of electricity reform among economic and social groups, and the impacts on unemployment. The other class of public benefits analyzed in this report—"environmental benefits"—received far less attention from policymakers, public interest groups, or Argentine consumers. The environmental issues that emerged in the reform process included the impact of the reformed tariff structure and unbundling on incentives for demand-side management or end-use efficiency, the development of significant national renewable energy resources, and meeting voluntary commitments to reduce greenhouse gas emissions. Interestingly enough, donors responding to international developments or the interests of their domestic constituencies were the most active proponents of preserving or enhancing environmental benefits.

Social Benefits: Creating Space for their Attention Post-Reform

Expanding Access to Isolated Rural Populations
A social issue that emerged in the reform process was the difficulty of expanding or maintaining basic access to electricity for the most isolated rural populations. Nationally, electrification access was 91 percent before reforms were implemented, and by

2000 had risen to 95 percent. But this improvement in access was due largely to the formalization of previously illegal connections in urban and suburban areas rather than to expansion of electricity services in rural areas. The majority of the populations that remain unconnected are located in isolated areas where it is costly to expand transmission or distribution grids—roughly 30 percent of the total rural population (Secretaría de Energía y Minería, 2001b). After privatization, it remained commercially unattractive for private distribution companies to connect these populations. Under the federal scheme, regulators cap the maximum price a distributor can charge end-users. Distributors maximize their returns by delivering electricity at the lowest average cost per kilowatt. Not surprisingly, most distributors made investments or improvements to distribution networks in densely populated urban and suburban areas. In these areas, per capita increases in demand and densification (more customers per square kilometer) reduced the average cost of delivering electricity, and thus allowed distribution companies to increase profits.

In 1995, the Secretary of Energy responded to the problem of under-served rural areas by announcing a five-year program known as Power Supply for Dispersed Rural Populations (PAEPRA). The program was designed to establish concession contracts for distributed energy to isolated populations (Secretaría de Energía, 1999c). Yet, by 1999, the PAEPRA program was falling far short of its announced goals despite complementary funding from the Global Environment Facility (GEF) and the World Bank to introduce a renewable energy component known as Renewable Energy for Rural Electricity Markets (PERMER). The original objectives of the PAEPRA project included connecting 314,000 households (1.4 million people), and 6,000 public services (schools, libraries, hospitals) in 16 provinces. At the end of 2000, EDJESA—an Argentine/Chilean company who was the sole concession in operation in the Province of Jujuy—had connected 3,107 rural customers (Secretaría de Energía y Minería, 2001a).

The principal problems encountered by PAEPRA included provincial governments' unwillingness to contribute subsidy payments to concession holders, as well as a lack of interest from commercial entities or companies. With support from the World Bank and the GEF, the program is being redesigned, yet its future is uncertain given the current national economic crisis and the pressure to reduce fiscal expenditures (Secretaría de Energía y Minería, 2001b). What is clear from this experience, however, is that federal efforts to provide rural electricity services require greater subsidies, and concession designs that go beyond tweaking the existing models operating in urban and suburban areas with distribution grids.

Connecting Poor Urban Households
In urban areas, the privatization of distribution services also affected the lowest-income consumers. Those least able to pay—illegally connected *"colgados"* ("hangers") concentrated in urban slums—were initially cut off from service by distribution companies. The International Finance Corporation (IFC) provided loans to distribution companies to fund technical and infrastructure changes that made electricity theft very difficult. As a result, non-technical losses in distribution networks (before privatization, losses reached 27 percent of the electricity supply) were drastically reduced (Inter-American Development Bank, 1995). These changes ensured the solvency of the distribution companies, but ignored the problem of how to provide basic electricity services to those without the economic resources.

A significant social scandal ensued over the termination of service to the *colgados*, and several court cases were brought against distribution companies. The basis for these cases was that privatization deprived a very significant population of basic services, even though they were obtained illegally. In response to negative media attention and mounting public pressure,[9] the federal government, the government of the Province of Buenos Aires, and two private distribution companies (EDENOR and EDESUR) entered into an agreement called the "Four-Year Framework Agreement" (Chisari and Estanche, 1999).

As part of this agreement, the federal government, the Buenos Aires provincial government, and municipalities reimbursed the companies for the unpaid balances associated with illegally connected shantytowns and provided subsidies to distributors to cover the cost of establishing collective meters and connections to their networks. In turn, the companies agreed to waive any claims, surcharges, or interest that had accrued on unpaid bills since 1992, pledged to install at least 10,000 meters a month in low-income areas, and agreed to conduct a household census with the informed consent of residents. As a result of this agreement, roughly 650,000 users were formally connected to the network (Chisari and Estache, 1999).

Federal efforts to provide rural electricity services require concession designs that go beyond tweaking the existing models operating in urban and suburban areas.

For the most part, consumption by low-income communities unable to pay for electricity is paid by municipal governments. Cities generally recoup these expenditures by imposing a tax on household electricity consumption. The public's unwillingness to accept the elimination of services to low-income populations was the primary reason the governments, companies, and municipalities negotiated a settlement that did not abide by the commercial principles established for distribution concessions.

Initial federal reforms retained some cross-subsidies. Decree 1398/1992 established subsidies for pensioners, public interest institutions, nonprofit organizations, and electricity-intensive industries. The National Electricity Act (Law 24065) established a National Electricity Fund capitalized from a tax on electricity sales in the wholesale market. Sixty percent of these tax revenues are distributed to provinces that adhere to the federal scheme for distribution tariffs to subsidize consumers; the remaining 40 percent is directed to electricity development in the country's

interior. Despite the retention of these cross-subsidies, there was a significant drop in levels of social tariffs enjoyed by lower-income groups post-reform.

The Impacts of Electricity Pricing on Low-Income Groups

Another relevant public benefits issue is the impact of electricity prices at the household level. By 1995, electricity prices in the wholesale market had fallen by more than 50 percent. This fact is widely cited and receives considerable praise from donors, economists, and energy specialists (ENRE, 1998; Green and Rodriguez-Pardina, 1999). But almost no mention is made of how these price declines were distributed among income or consumption classes. A recent study on the effect of reforms on prices indicates that between 1991 and December 1998, residential and industrial consumers with the highest levels of electricity consumption experienced the largest price declines (71 percent and 44 percent respectively). Households with low consumption levels, generally representing lower-income populations, experienced only marginal price declines (1.6 percent) (FLACSO/SECYT-CONICET, 1999). In general, industrial concerns, large consumers (2 megawatts-50 kilowatts or more per annum), and distribution companies that could buy electricity on the wholesale market experienced price declines of 50 percent or more. Absolute and relative prices by consumption classes are shown in Table 3.1.

Several aspects of the reform produced regressive social pricing. The new regulatory framework required that electricity services reflect the cost of supply. Distribution costs are in inverse proportion to the quantity and voltage of the supply. Thus, consumers with low consumption and voltage levels paid more relative to industrial and high residential consumers. In effect, the more one consumed, the lower the per-unit price paid. The discrepancy in prices may also be due to a possible error on the part of ENRE—in its allocation of distribution costs to be covered by low- versus high-consumption households—that remained uncorrected during the first 10-year tariff period.[10]

Interestingly, in 2000, ENRE initiated the first 10-year distribution tariff revision process, and included requirements to incorporate social tariffs and market incentives for energy efficiency in the development of the next 10-year price cap. Unfortunately, the tariff revision process was suspended in 2002 with the dissolution of the pesos' peg to the U.S. dollar and the Euro.

Another contributing factor was the five-year fixed-price electricity supply contracts that were bundled with the distribution concessions when they were initially privatized. These long-term price contracts were designed to reduce the risk of price fluctuations faced by potential private investors. The contracts covered up to 50 percent of projected demand for electricity within a concession area. As a result, although prices in the wholesale spot market where distribution companies purchased the remainder of the electricity needed to meet demand dropped dramatically, much of these savings were not passed on to residential consumers.

By the late 1990s, World Bank staff in the Argentina country office recognized that power sector reforms had produced regressive economic and social effects. Some of the conclusions arrived at in a World Bank seminar on Argentina's public sector reforms included the following:

- The poor were most affected by the privatization process;
- The rich were the greatest beneficiaries of the privatization process;
- The system should internalize transfer mechanisms favoring the poor (cross-subsidies);
- A social tariff should be established based on the capacity-to-pay; and
- New concessions should include a clear strategy to force operators to serve sectors with limited economic resources (World Bank, 1999).

Early in the reform process, the World Bank argued against cross-subsidies and for the elimination of social tariffs because they would distort price signals and weaken budgetary discipline (World Bank, 1993).

<table>
<tr><td>TABLE 3.1</td><td colspan="3">ELECTRICITY PRICES IN ARGENTINA, 1991–1998</td></tr>
</table>

US cents per kilowatt-hour	March 1991	December 1998 (compared to March 1991= 100)
Residential low consumption	8.2	8.1 (98.4 %)
Residential high consumption	15.9	4.7 (29.6 %)
Industrial low consumption	14.0	10.5 (75.3 %)
Industrial average	8.4	7.4 (88.5 %)
Industrial high consumption	5.6	3.7 (66.6 %)
Average	8.8	7.8 (89.1 %)

Source: FLACSO/SECYT-CONICET. 1999. *"Privatizaciones en la Argentina: Regulación tarifaria, mutaciones en los precios relativos, rentas extraordinarias y concentración económica."* Buenos Aires, April.

The World Bank and other donors also argued that efficiency improvements generated by market forces would be passed on to consumers in the form of lower prices, particularly for the poor (World Bank, 1993). Ironically, by the end of the 1990s, the World Bank was arguing for increases in social tariffs and actions to correct the market's failure to serve income groups with limited resources.

Electricity Privatization and Unemployment
During the period in which the Argentine government implemented its public sector reform program, Argentina's unemployment rate climbed from 6.8 percent in 1991 to 20 percent in 1995, and fluctuated between 14-18 percent for the period 1996-99. Government privatizations displaced an estimated 350,000 individuals through the loss of public sector jobs or the elimination of jobs that depended on the public sector (Clarín, 2000; Pistonesi, 2000).

In the electricity sector, employment fell from 22,500 before the reform process began to 6,500 in 1998 (Duarte, 2001). More than 50 percent of these losses were sustained between 1991 and 1993. They came about through dismissals, early retirement plans, voluntary retirement programs, and hiring freezes. Financing from the World Bank supported early and voluntary retirement packages in the 1991-93 period (Duarte, 2001). Above and beyond voluntary retirement packages, however, the government did not put in place any comprehensive policies to address unemployment impacts at the time that it privatized public sector enterprises. In fact, governments and donors only began to implement job creation and labor policy reforms in the late 1990s, when unemployment had turned into a persistent and chronic problem.

Most efforts to address or enhance environmental benefits did not enjoy the political support of senior policymakers. Most striking is the lack of demonstrated interest of Argentine public interest environmental groups.

After 1993, the newly privatized distribution companies undertook an additional wave of layoffs. These layoffs were not voluntary, but were accompanied by generous retirement or severance packages. After they reduced their labor pool, distribution companies ensured adequate staffing by outsourcing functions or services previously carried out by in-house employees to third party contractors. Third-party contracting allowed distribution companies to obtain services at lower cost (contract employees enjoyed fewer benefits, lower pay), and on an as-needed basis (Pistonesi, 2000). Although many former utility employees were re-hired by third party contractors, the terms of their employment were generally less favorable, as was the case nationally for most of the unemployed (Martínez, 1998).

Environmental Benefits: A Case of Limited Political and Public Interest

In Argentina, concerns about how reforms affected the environment and where opportunities existed to generate environmental benefits received little or no attention from national policymakers or environmental organizations. Most efforts to address or enhance environmental benefits were isolated or did not enjoy the political support of senior policymakers. Most striking, however, is the lack of demonstrated interest, with a few exceptions, of Argentine public interest environmental groups.

Establishing Environmental Regulation in the Electricity Sector

In the early 1990s, the main concern of the reformers was to update and modernize environmental regulations in the electricity sector. ENRE was given authority to set environmental rules and regulations and to outline the procedures necessary for generation, transmission, and distribution to comply with these regulations. Procedures were spelled out in environmental annexes (dating from the 1980s) that established environmental impact assessment procedures for thermal generation, hydroelectricity, and transmission projects. The annexes were expanded to include management plans for compliance with ambient air quality and emission standards, and requirements for periodic auditing and monitoring by ENRE.

Among the banks, the IDB made a conscious commitment to strengthen the government's capacity to develop, regulate, and enforce environmental laws and policies. This commitment included loans for capacity building and strengthening for Argentina's Secretary of Environment and Natural Resources and the Secretary of Energy's social and environmental assessment unit.[11] There were few if any actions taken beyond these IDB-funded efforts to modernize environmental regulation. The main reason for this was that the IDB, the World Bank, and the federal government initially shared the view that instituting market reforms would establish price signals that reflect scarcity values, drive improvements in efficiency, and ultimately encourage the rational use of energy resources.[12] Beginning in the mid-to-late 1990s, however, donors showed their growing interest in addressing environmental problems by increasing funding to projects that examined the environmental dimensions of energy consumption and development.

Improvements in Energy Efficiency
Some of the efficiency gains predicted by the federal government materialized. For example, transmission and distribution losses fell from an all time high of 27 percent to below 10 percent for the period 1992-97 (ENRE, 1998). Most of these reductions were achieved by eliminating illegal connections to the grid through grants provided by donors or cross-subsidies imposed by municipal governments. In addition, thermal generation of electricity increased in efficiency as natural gas plants displaced older thermal plants and reduced the energy consumption per kilowatt-hour from 2,600 kilocalories in 1996 to 1,600 kilocalories in 2000 (Vásquez, 2000). Nevertheless, the efficiency improvements in natural gas generation must be examined in the context of other power generation sources. The energy and carbon intensity of electricity production in Argentina declined between 1986 and 1996, but this is largely due to new capacity additions from hydropower projects initiated before the electricity reform program (Olander, 2000). Given the significant expansion in natural gas plants slated to come on-line on or before 2004, the emissions intensity of electricity generation is actually projected to increase from 17.9 metric tons in 1998 to 36.5 metric tons by 2010 (Olander, 2000; Secretaría de Energía, 1999b).

Managing Electricity Demand
Because the advantages of energy efficiency are captured primarily in generation, vertical unbundling reduces the incentives at the distribution level to increase efficiencies among end users or final consumers. Even so, the high incidence of fixed costs faced by distributors should encourage them to improve network load factors in order to postpone new investments. But, in Argentina, the continual downward revision of what defined a large electricity consumer (lowered to 50 kilowatts per annum in 1998) eroded any remaining incentives distributors might have had to invest in energy efficiency. Distributors were left with a customer base of low-

consumption households and businesses, lengthening the payback period and reducing the potential savings from such investments. Among residential households, the highest income consumers accounted for the greatest increases in energy consumption. This group experienced falling per-unit prices because of the decreasing block price structure, and consequently had few incentives to invest in energy efficiency. *(See Table 3.2.)*

Attempts by donors to promote energy savings companies (ESCOs) or engage distribution companies in efforts to get consumers to reduce or manage their energy consumption have been short-lived. Chief among these was the IDB's attempt to establish a project—dubbed sustainable markets for sustainable energy (SMSE) —that would, among other things, promote ESCOs and energy-efficiency services in Argentina. A lack of official interest and changes in personnel in the Secretary of Energy, led the IDB to end the program in Argentina.[13]

Neither the creation of distributed power concessions nor the subsidies offered by international donors proved sufficient to draw investment to renewable energy for distributed power.

Although the Secretary of Energy had a National Office for the Rational Use of Energy (URE), it did not promote comprehensive policy instruments or efforts. URE was largely devoted to implementing a suite of donor-funded pilot projects and programs, particularly from European Union (EU) agencies and governments.[14] Outside of URE, it was also donors that demonstrated the greatest interest in addressing energy-related environmental problems. Some of these donor-funded efforts included: the transfer of energy-efficiency technologies and practices to small and medium enterprises (GTZ); demand-side management and integrated resource planning (USAID); and the renewable energy component of the PAEPRA program (GEF and World Bank). Although these

programs were initiated in collaboration with the
Secretary of Energy, most were not institutionalized
(Secretaría de Energía y Minería, 2001b).

Development of Alternative and Renewable Generation Sources

Argentina has abundant small hydro and wind
resources, but these have attracted little or no
investment because falling electricity prices—from
the development of natural gas resources and the
return of normal hydroelectricity production after the
drought of the early 1990s—limited their commercial viability. For the better part of the 1990s, private
investments generally went to upgrading or building
gas-fired power plants and developing natural gas
fields and pipelines. As of 1999, investments in
natural gas plants, exploitation, and transportation
were valued at $1.5 billion in Argentina, and $2
billion in binational projects with other countries
(largely Chile and Paraguay) (Campodónico, 1999).
Neither the creation of distributed power concessions
under the PAEPRA program nor the subsidies
offered by international donors under the PERMER
project proved sufficient to draw investment to
renewable energy for distributed power.

The lone example of a domestic effort to support
renewable energy was a 1998 campaign led by
Greenpeace to promote wind power in southern
Argentina. This campaign was funded heavily by
wind-power developers.[15] The resulting public support
generated by the campaign led Argentina's congress to
approve legislation that provided a subsidy to companies that manufacture wind turbines domestically.
This legislation had little buy-in from the executive
branch, and the Secretary of Energy delayed issuing
supporting regulations until December 1999. Because
of this delay and the failure of wind developers to
build turbines domestically, very few if any projects
have taken advantage of this legislation.

Early in 2001, ELECNOR and ENDESA —an
Argentine distribution company and Spanish energy
corporation respectively—announced their intention
to form a renewable energy joint venture, Energías
Argentinas S.A. (ENARSA). They committed to
building 3,000 megawatts of wind power generation
in four southern provinces. The total estimated
investment at the time was $2.3 billion over 10 years
(Office of Argentine President, 2001). Interestingly
enough, the legislation that was passed as a result of
the Greenpeace campaign appears not to have
motivated the formation of this joint venture. In fact,
ELECNOR and ENDESA qualified their commitment
by noting that it would depend on the construction of
local distribution networks and a supportive regulatory regime that ensured price stability over a 15-year
period. The price stability and investments in
infrastructure called for by ENARSA now seem
highly improbable, given Argentina's debt default
and generalized economic crisis.

Following Through on Climate Change Commitments

Climate change analysis and policy follows a trajectory similar to efforts to address the environmental
dimensions of energy. In 1997, donor funding and
interest, particularly from the U.S. government, led
the Secretary of Energy to undertake analyses that
permitted Argentina to formulate a voluntary
national commitment to reduce greenhouse gas
emissions. The commitment was announced at the
Fifth Conference of the Parties of the United Nations
Framework on Climate Change (UNFCCC) hosted by
Argentina in November 1998. After the fact, neither
the Menem administration that was responsible for
making the commitment, nor the subsequent but
short-lived de la Rua administration, ever gave any
official indication of whether or how they planned to
fulfill it. Even before the current crisis, which has
focused national attention on economic survival,
there was extremely limited public interest or
awareness in this commitment or in climate change.

MANAGING MARKETS IN THE PUBLIC INTEREST: THE STATE'S REGULATORY ROLE

Framing the above discussion is the larger question
of the role the state should assume in an economy
where important public services are provided

through competitive markets. The Menem government's initial concern in the reform process was to establish a strong market regime that would attract and retain private sector participation. This concern is reflected in the official mandate of ENRE, the electricity regulator: mediation of disputes between electricity companies; enforcement of federal laws and regulations and terms of concessions; definition of service standards for distribution companies; and determination of the maximum prices transmission and distribution companies may charge (Estache and Rodriguez-Pardina, 1996). Defense of the public interest is implicit in this document, but it is unclear how and when it is appropriate for ENRE to defend consumers or the public interest.

One incident illustrates ENRE's cautious efforts in defense of consumer interests. The incident in question was a February 1999 blackout that affected 160,000 residential customers. Known as the "EDESUR incident," it knocked out telephones, water supply, traffic lights, and subway lines in a central Buenos Aires neighborhood for upwards of 10 days. Although this incident might constitute an isolated event, the distribution company—EDESUR—had a history of relatively poor performance, and accounted for over half of all consumer complaints related to distribution services in the Greater Buenos Aires metropolitan area for the period 1993–99 (ENRE, 1998; 2000). During the remainder of 1999, additional problems with distribution services plagued the Buenos Aires metropolitan region, prompting ENRE to call meetings with distribution companies to investigate the reason for blackouts and consumer complaints (*Petrochemical, Petroleum, Gas and Chemical Magazine, 1999*).

During the EDESUR incident, ENRE very narrowly defined its role—to punish the distributor and to satisfy consumer demands that service be re-established. Despite evidence of poor performance and repeated problems with distribution services, ENRE focused largely on retroactive corrections. Later in 1999, ENRE belatedly brought together distribution companies to discuss larger service quality and

reliability issues arising from the EDESUR incident and other repeated blackouts.

During the same period, the capacity to ensure optimal investments in transmission expansion, another ENRE responsibility, also came into question. Since the mid-1990s, parts of Argentina's national interconnected grid have been running at or near full capacity, increasing the risk of large-scale power outages. Despite this fact, the only two transmission expansion proposals to come before ENRE in the mid-1990s were blocked by beneficiaries that rejected the proposed transmission charges (Abdala and Chambouleyron, 1999). Private contracts for individual lines linking particular industrial customers or communities to the grid were approved by ENRE, but these did not solve the larger congestion problems plaguing the 500-kilovolt lines that interconnect the national grid (Abdala and Chambouleyron, 1999).

In the final months of the second Menem administration (December 1999), the Energy Secretary, César Mac Karthy issued an executive order establishing a Federal Electricity Transmission Fund (FETF). The fund was to be capitalized with a surcharge of US$0.06 per megawatt on large consumer and distribution company purchases in the wholesale electricity market. In principle, the FETF was created to finance in whole or part the construction of high-tension (500 kilovolt) transmission lines identified as a priority by the Federal Electricity Commission (Secretaría de Energía, 1999a). Mac Karthy's order responded to an increasing number of blackouts in late 1999—over 80 in the last 50 days of that year. Many were linked to the fact that electricity supply and demand were exceeding the capacity of the transmission system *(Petrochemical, Petroleum, Gas and Chemical Magazine, 1999)*.

Fernando de la Rua's administration, which took office in January 2000, supported a renewed federal role in the provision of transmission infrastructure, and even went as far as to question the soundness of privatization and the regulatory system's capacity to ensure reliable electricity services. President de la Rua called for "a thorough investigation of the design of privatization to determine what problems exist from a technical perspective and identify whether there are guarantees for service provision." *(Petrochemical, Petroleum, Gas and Chemical Magazine, 1999)*. Soon after this statement, the de la Rua administration was caught up in a larger economic crisis that took precedence over electricity concerns.

Desperate to find a way out of Argentina's deep recession and looming $140 billion debt obligations, de la Rua invited Domingo Cavallo to become Economy Minister in March 2001 and to put in place an emergency economic recovery plan (Calero, 2001). Cavallo pushed through labor reforms, deregulation of healthcare and telecommunications, budget cuts of $1.4 billion, tax increases of $2 billion, and public sector wage cuts of 15 percent (Colitt, 2000). This was followed in August 2001 by additional austerity measures (including a zero deficit budget), reduced income and consumption taxes to stimulate growth, and negotiations of emergency loan packages with the International Monetary Fund (IMF) and the IDB (BBC News, 2001; Drajem, 2001).

Cavallo, and the team he brought with him, reasserted the need for market discipline and significant public sector reforms. Cabinet ministries were reorganized, and Carlos Bastos—the chief architect of the reform and privatization of the electricity sector in the early 1990s—was named Minister of Infrastructure and Housing, which included the Secretary of Energy and Mining. In June 2001, Bastos suspended Mac Karthy's executive order establishing FETF, and issued a separate decree that reaffirmed the original electricity reforms. The decree introduced a new market instrument (congestion licenses) intended to make invest-

ments in transmission more attractive. Additionally, it established a transmission reimbursement fund to provide additional payments to transmission companies, parties to BOM contracts, or holders of congestion licenses, if and when their investments enhanced the overall stability of the transmission system (Poder Nacional Ejecutivo, 2001). The latter half of 2001 saw few additional changes to electricity policy or regulation. The economic crisis engulfing Argentina led Cavallo and de la Rua to resign in December 2001 amid violent street protests.

The current crisis is leading Argentine policymakers and the public to reassess the country's opening to foreign investment, and the economic constraints dictated by repeated IMF, World Bank, and IDB bailouts. To maintain the country's access to private and multilateral development bank credits, de la Rua and Cavallo continued to cut social expenditure and wages in the hopes of boosting productivity and releasing revenues to service an ever-growing and more onerous foreign debt (Felix, 2001). The impact on Argentina's electricity sector is potentially serious. Consumers will pay in devalued pesos or not pay at all. Electricity companies (distribution, transmission, and generation) that hold debts in foreign denominated dollars may see their revenues shrink and be forced into bankruptcy or default. Shareholders in these companies will see the nominal value of their assets disappear. A difficult road lies ahead for Argentina's electricity sector, economy, and society.

CONCLUSIONS

This analysis of Argentina's reforms does not point to a particular path that developing countries should follow to preserve public benefits in a deregulated electricity sector. However, the analysis does permit the following conclusions about the realities of defending public benefits in Argentina:

- The economic crisis of the late 1980s and early 1990s permitted policymakers and reformers to revisit the whole structure of the electricity sector. Such a revisiting may occur again, given the country's current crisis. Although this might bode well for public benefits, past experience also indicates that it is extremely difficult to focus policymakers and donors on the issue, which is perceived as a second order objective under such circumstances.

- Argentina's experience during the 1990s demonstrates that public benefits do not flow automatically from a financially solvent and efficient electricity sector. Public benefits require explicit attention, and there is a greater likelihood they can be enhanced if they are considered when reforms are designed and first implemented. After reforms are implemented, it is much harder to introduce changes that will favor social and environmental benefits.

- When the public at large demonstrated a direct interest in a public benefit, it was possible to broker solutions (e.g., the EDESUR incident or the termination of service to poor urban slums) that preserved or enhanced its supply. In the absence of such interest (for example, climate change), minority or expert constituencies rarely exercised any leverage or ability to influence the terms of the public debate or to change public policies.

- A few donors, particularly the World Bank, had considerable influence at the *initial* stage of Argentine reforms, because at that time the Argentine government's ability to carry forward its reform plans depended on obtaining sufficient resources from donors. After this initial period, however, most donor leverage was limited to tinkering around the edges of the restructured electricity sector.

- A narrow definition of the state's regulatory mandate focused on the enforcement of market rules limits the state's leverage to prospectively and pro actively incorporate public benefits into regulatory activities. Electricity regulation in Argentina requires a broader mandate that explicitly balances the regulator's responsibility to ensure that markets function with the need to prevent encroachment on public interests.

NOTES

1. Interview with IDB staff person, August 10, 2000. All interviews for this chapter were conducted on a not-for-attribution basis. Consequently, interviewees are identified only by their institutional affiliation.

2. Interview with former Energy Secretary staff and academic energy experts, Spring 2000.

3. Interview with former staff at the Secretary of Energy, Spring 2000.

4. Interview with World Bank staff in Latin American and Caribbean Department, September 22, 2000.

5. Interview with IDB staff, July 27, 2000.

6. Interview with IDB staff, July 27, 2000.

7. Interview with IDB staff, July 27, 2000 and interview with World Bank staff in the Latin American and Caribbean Department, September 22, 2000.

8. Interview with World Bank staff, September 22, 2000 and interview with IDB staff, July 22, 2000.

9. Interview with former staff at the Secretary of Energy, Spring 2000.

10. Interview with IDB staff, March 14, 2002.

11. Interview with IDB staff, July 22, 2000.

12. Interview with World Bank staff, September 22, 2000 and interview with IDB staff, July 22, 2000.

13. Interview with IDB staff, August 3, 2000.

14. Interview with IDB staff, August 3, 2000.

15. Interview with a commercial wind-power developer, Spring 2000.

REFERENCES

Abdala, Manuel A. 2001. "Institutions, Contracts and Regulation of Infrastructure in Argentina." *Journal of Economic Literature*. IV (November): 217-254.

Abdala, Manuel A. and Andres Chambouleyron. 1999. "Transmission Investment in Competitive Power Systems: Decentralizing Decisions in Argentina." Public Policy for the Private Sector Note No. 192. Washington, D.C.: World Bank. (September).

Bastos, Carlos M. and Manuel A. Abdala. 1996. *Reform of the Electric Power Sector in Argentina*. Buenos Aires.

BBC News. 2001. "Argentina Minister Appeals for Calm." Business section. July 13. Online at: http://news.bbc.co.uk/hi/english/business/newsid_1436000/1436129.stm (May 1, 2002).

Besant-Jones, John. 1996. "The England and Wales Electricity Model–Option or Warning for Developing Countries?" Public Policy for the Private Sector Note No. 84. Washington, D.C.: World Bank. (June).

Calero, Róger. 2001. "Economic Crisis Spurs Protests Across Latin America." *The Militant*. 65(22).

Campodónico, Humberto. 1999. "The Natural Gas Industry and its Regulation in Latin America." *CEPAL Review* No. 68 (August). Santiago: United Nations Economic Commission for Latin America and the Caribbean.

Chisari, Omar and Antonio Estache. 1999. "Universal Service Obligations in Utility Concession Contracts and the Needs of the Poor in Argentina's Privatizations." WPS 2250. Washington, D.C. : World Bank. Online at: http://www.worldbank.org/wbi/regulation/pdfs/ (May 1, 2002).

Clarín. 2000. "Privatizaciones: las asignaturas pendientes." Tapa suplemento Económico. (August 13). Online at: http://www.clain.com/suplementos/economico/2000/08/13/index.html (May 1, 2002).

Colitt, Raymond. 2000. "Uphill Battle for Argentina." *Financial Times*. (September 20). Online at: http://specials.ft.com/worldeconomy2000/FT33VSTPBDC.html (May 1, 2002).

Drajem, Mark. 2001. "Argentina's Cavallo Says He Thanked U.S., IMF During Visit." *Bloomberg Latin America*. (August 29).

Duarte, Marisa. 2001. "El impacto del proceso de privatizaciones sobre el empleo de las empresas prestatarias de servicios publicos." Buenos Aires: FLACSO – Area de Economía y Tecnología.

Entidad Nacional de Regulación Eléctrica (ENRE). 1998. *Informe Eléctrico: 5 años de regulación y control. 1993-Abril 1998*. Buenos Aires.

———. 2000. *Annual Report for 1999*. Buenos Aires.

Estache, Antonio and Martin Rodriguez-Pardina. 1996. "Regulatory Lessons from Argentina' s Power Concessions." Public Policy for the Private Sector Note No. 92. Washington, D.C.: World Bank. (September).

———. 1999. "Light and Lighting at the End of the Public Tunnel: Reform of the Electricity Sector in the Southern Cone." Regulatory Reform and Private Enterprise Division. Economic Development Institute. Policy Research Working Paper No. 2074. (March).

Felix, David. 2001. "Is Argentina the *Coup de Grace* of the IMF's Flawed Policy Mission?" *Foreign Policy in Focus*. Interhemispheric Resource Center and the Institute for Policy Studies. (November). Online at: http://www.foreignpolicy-infocus.org/pdf/reports/argentina.pdf (May 1, 2002).

Green, Richard and Martin Rodriguez-Pardina. 1999. "Resetting Price Controls for Privatized Utilities: A Manual for Regulators." Washington, D.C.: World Bank.

FLACSO/SECYT-CONICET. 1999. "Privatizaciones en la Argentina. Regulación tarifaria, mutaciones en los precios relativos, rentas extraordinarias y concentración económica." Buenos Aires: (April).

Hasson, Graciela. 1994. "Análisis de la Privatizaciones Eléctricas." *Desarrollo y Energía*. 3 (5).

Inter-American Development Bank. 1995. Project Summary for Project No. AR-0195 (Edenor S.A.). Washington, D.C.

———. 2000. *Energy Sector Strategy*. Annex Table 1. Sustainable Development Department. Sector Policy and Strategy Paper Series. (May).

Margheritis, Ana. 1999. "Características e impacto de la implementación del programa de privatizaciones en la Argentina." Baima de Borri, Marta et al. (comp.) *La privatización de los servicios básicos y su impacto en los sectores populares en Argentina*. Buenos Aires: Banco Mundial, Grupo de Trabajo de ONGs Sobre el Banco Mundial, and Instituto de Investigaciones del Nuevo Estado. Editorial de Belgrano.

Martínez Gerardo. 1998. "The Prospect of Reducing Poverty and Improving Social Balance in Latin America: The Improvement of Employment Opportunities." Conference paper. Fourth Annual World Bank Conference on Development in Latin America and the Caribbean, San Salvador, El Salvador, June 28-30.

Poder Nacional Ejecutivo. 2001. "Decree 0804/2001." Published in Official Bulletin No. 29.673. Buenos Aires. (June 21).

Office of the Argentine President. 2001. "Invertirán 2,250 millones de dólares para construir parques eólicos." Press Release. (February 6).

Olander, Jacob. 2000. "Climate Change, Private Investment and Power Generation in Latin America." Unpublished study produced for the World Resources Institute. (December).

Petrochemical, Petroleum, Gas and Chemical Magazine. 1999. "El Año Que Pasó: Diagramas en Tensión." Vol. 158 (December).

Pistonesi, H. 2000. "The Argentine Electricity Sector: Post-Reform Performance." Institute for Energy Economics. Buenos Aires: Fundación Bariloche.

Secretaría de Energía. 1999a. Resolution SE No. 0657/1999. Published in the Official Bulletin. No. 29.292. Buenos Aires. (December 14).

———. 1999b. *Prospectiva 1999.* Buenos Aires.

———. 1999c. "Abastecimiento Eléctrico a la Población Rural Dispersa Argentina." Online at http://energia.mecon.gov.ar (August 22, 2001).

Secretaría de Energía y Minería. 2001a. *Informe del Sector Electrico Año 2000.* Buenos Aires. (October).

———. 2001b. *Prospectiva 2000.* Buenos Aires. (April).

———. 2001c. "Proyecto de Energías Renovables en Mercados Rurales." Online at: http://energia.mecon.gov.ar/electricidad/permer/default.htm (May 10, 2002).

Vásquez, Patricia I. 2000. "El Sector Eléctrico en Argentina: Reconversión y Mitigación de Gases de Efecto Invernadero." Unpublished study produced for the World Resources Institute. (April).

World Bank and Organización Latinamericana de Energía (OLADE). 1991. *"The Evolution, Situation, and Prospects of the Electric Power Sector in the Latin American and Caribbean Countries, Vol. II. Descriptions of Individual Power Sectors."* Latin America and the Caribbean Technical Department, Regional Studies Program, Report No. 7. (August).

World Bank. 1993. "The World Bank's Role in the Electric Power Sector: Policies for Effective Institutional, Regulatory and Financial Reform." Policy paper 11676. Washington, D.C.: World Bank.

———. 1995. "Privatizing Argentina's Public Enterprises." Operations Evaluation Department. Precis No. 100. Washington, D.C. : World Bank. (December 1).

———. 1996. "Argentina: Reforma de las empresas provinciales de servicios públicos: Problemas, desafíos y mejores prácticas." Buenos Aires: World Bank. (June).

———. 1999. "The Privatization of Basic Services and its Impact on Low-Income Sectors in Argentina." Seminar Report. Buenos Aires. (May).

———. 2000. Operations Evaluation Department Memorandum to the Executive Directors and the President. Subject: Argentina Country Assistance Evaluation. (July 10).

4

INDIA

ELECTRICITY REFORM UNDER POLITICAL CONSTRAINTS[1]

Navroz K. Dubash
Sudhir C. Rajan

INTRODUCTION

For much of the history of post-independence India, the electricity sector has been an entrenched symbol of the nation's state-led economic development approach. Publicly owned, and operated and managed by state employees, the sector was conceived of and run as an instrument of development policy. Beginning in 1991, however, these basic assumptions began to be challenged. Sector reform efforts have been as much about contesting this mindset as about undertaking changes in ownership, investment, and management practices. For a profile of the electricity sector in India see Box 4.1.

Electricity sector reform in India has become polarized. Efforts to shrink the role of the state and replace it with greater private sector participation allowed little or no place for state stewardship of a public benefits agenda. On the other hand, efforts to continue operating the sector as an instrument of development policy failed to recognize the dire state of the sector. This study of the political economy of decisionmaking seeks to go beyond this dichotomy to understand how public benefits can be promoted in a post-reform sector. A central theme of the chapter is the need for more democratic decisionmaking in the sector.

There have been four overlapping but distinct periods of electricity sector policy approaches: (1) pre-1991; (2) the 1991 indepedent power producer (IPP) policy and its aftermath; (3) the World Bank-led

BOX 4.1	PROFILE OF THE ELECTRICITY SECTOR IN INDIA

Population (2001)[1] : 1.0 billion.

Population with access to electricity (2000)[2]:
Total: 46%   *Rural:* 33%   *Urban:* 82%

Installed electricity generation capacity (1999)[3]
Total: 103 gigawatts (3.2% of total world capacity)
> *Thermal:* 76%
> *Hydro:* 21%
> *Nuclear:* 2%
> *Geothermal and Other:* 1%

CO_2 emissions from electricity and heat as a share of national emissions (1999)[4]: 53%

Notes:

1. World Resources Institute. 2000. *People and Ecosystems: The Fraying Web of Life.* Washington, D.C.: World Resources Institute.

2. International Energy Agency. 2002. *Electricity in India: Providing Power for the Millions.* Paris.

3. www.eia.doe.gov/pub/international/ieapdf/to6_04.pdf (February 6, 2002).

4. Computed by WRI using International Energy Agency (IEA) data. IEA, 2001. *CO2 Emissions from Fossil Fuel Combustion.* Paris: OECD.

restructuring policy, which began to be implemented around 1993 in Orissa; and (4) the period shortly after 1998, when the restructuring model was scaled up through national legislation and state-level reforms. In this report, these periods are described thematically rather than sequentially. Nonetheless, distinguishing between them is useful in order to recognize how and when different types of institutional arrangements were "locked in" with considerable impact on the electricity sector.

BACKGROUND: A LEGACY OF STATE CONTROL

During the 1990s, electricity sector reforms were part of a seismic shift in India from a closed toward a more open economy. From Indian independence in 1947 until the mid-1980s, the state played a strong role in planning and implementing strategies for economic development. Internal and external pressures to rethink this approach emerged in the 1980s, as the country went through a moderate recession. These views were endorsed primarily by strong statements from development agencies that their borrowers would henceforward have to increasingly look to international capital markets for their financing needs.[2]

The immediate impetus for action was a serious balance of payments crisis in 1991. The response was to liberalize investment in key sectors of the economy, including electricity, to reduce licensing restrictions on industry, lift government controls on the financial sector, and partially free currency transactions. Both the intent, and the actual policies, marked a significant departure from the previous 40 years of government policy.

The Electricity Sector Before 1991

Operating under the Electricity Act of 1910, private companies or local authorities supplied more than 80 percent of the total generation capacity in the country prior to independence in 1947 (World Bank, 1993b). In 1948, the Electricity Supply Act brought all new generation, transmission, and distribution facilities within the state's purview. Each state subsequently established its own vertically integrated state electricity board (SEB).[3] Significantly, SEBs were financed through state government loans and were run as extensions to state energy ministries.[4] As a result, SEBs were "indebted in perpetuity," and were forced to continue in a relationship of financial dependence and administrative thrall to energy ministries.[5] Nonetheless, SEBs were the backbone of the electricity infrastructure, and by 1991 controlled 70 percent of electricity generation and almost all distribution (World Bank, 1991).

Under the Indian constitution, the electricity sector is a "concurrent" subject, allowing both the central and state governments some authority in the sector. SEBs are under the control of state governments, which also controlled the critical tariff-setting function. The central government was responsible for electricity policy, long-term planning, technical analysis, and project approvals through the Power Ministry, Planning Commission, and Central Electricity Authority. *(See Figure 4.1.)*

In addition, in response to declining SEB performance and to establish a "model of modern operational practices that the SEBs could emulate," the central government established two central power generation corporations—the National Thermal Power Corporation (NTPC) and the National Hydroelectric Power Corporation (NHPC) (World Bank, 1999a).[6] NTPC, now the world's sixth largest thermal power company, is widely considered an efficient and well-respected public corporation.[7]

By 1991, the first four decades of public-sector-led electricity development had chalked up some notable accomplishments. Between 1948 and 1991, generation capacity increased by a factor of 50 with an annual growth rate of 9.2 percent—considerably greater than the economic growth rate (World Bank, 1991). Moreover, official reports claimed that electrification rates were 80 percent.[8]

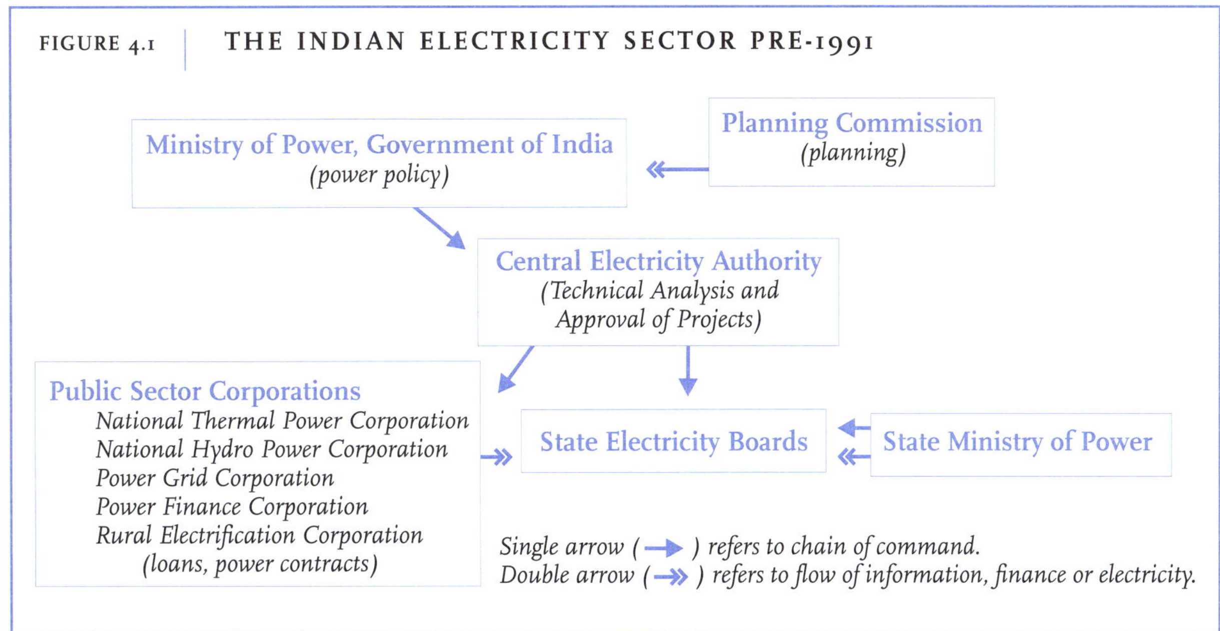

The Seeds of Crisis

Despite these accomplishments, there were reasons for concern about the future of the sector. Well before 1991, the sector had been locked into arrangements with electricity users, and into management practices with negative long-term implications. These arrangements constrained future reform efforts.

Perhaps the most damaging practice was the political decision in many states to provide highly subsidized or free electricity to farmers. Provision of electricity to run irrigation pumps was an important ingredient in the Green Revolution technology package aimed at increasing the productivity of Indian agriculture. However, from 1977 onward, electricity increasingly became an instrument of populist politics. By offering electricity at flat rates—based on pump capacity rather than metered consumption—or even completely free, several state governments cultivated farmers as a vote bloc.[9] Subsidized electricity imposed high costs and compounded the technological, institutional, and political problems in the sector.

These practices had several negative effects. First, by the mid-1990s, the World Bank estimated that

SEBs paid an annual subsidy of about $4.6 billion (1.5 percent of GDP) to agricultural and residential users (World Bank, 1999a). Second, since flat rate or free electricity rendered the meter redundant, existing meters were no longer monitored and were often broken or removed. This "de-metering" has increased the financial and organizational challenge to the re-introduction of a consumption-based tariff. Third, agricultural tariff policy has had negative spillover effects on overall management practices of the SEBs. Since electricity load for agriculture is not well measured, technical losses as well as theft throughout the sector are conveniently allocated to agricultural consumption (Reddy and Sumithra, 1997). Finally, although agricultural electricity subsidies have been introduced in the name of social benefits, poor farmers typically do not benefit from this subsidy, and indeed may be hurt by it.[10] However, wealthier farmers have successfully organized themselves to lobby for continuation of this policy.

Other negative effects followed. Although many states had a declared social policy to provide agricultural subsidies, they did not always pay the SEBs directly to compensate for the loss of revenue. Indeed, agricultural de-metering meant that the actual level of compensation required was often a mystery. Instead, SEBs developed an elaborate and self-defeating system of cross-subsidies from industrial consumers to make up for the growing revenue losses from agriculture and theft. Over time, industrial consumers found it more cost-effective to set up their own captive power plants to supplement, or replace, SEB electricity. In 1960, industrial consumption accounted for 67 percent of SEB sales; by 1991, its share had dwindled to 40 percent. Over the same period, agriculture consumption leaped from 10 to 25 percent (Tata Energy Research Institute, 1993). Losses from theft also seemed to be a serious problem. SEBs seemed reluctant to acknowledge the extent of such losses, perhaps because it was so difficult to distinguish theft from technical losses and unmetered consumption. Recent evidence suggests that while the focus has been on agricultural losses, industries using high-tension lines may be responsible for much of the theft and loss (Purkayastha, 2001; Mahalingam, 2002).

Hence, the SEBs found themselves in the unenviable position of facing growing loss-making segments of their business, and a shrinking profit-making segment. Considerable staff development and morale problems followed, with wages stagnant and sales per employee among the lowest in the world (Gutiérrez, 1993). The quality of the electricity provided inevitably suffered, with low frequency, brownouts and blackouts, and billing problems increasingly common. Poor service quality hastened the exit of industrial users from the grid, and diminished the willingness of consumers to accept higher tariffs, both of which accelerated the spiral of deterioration.

Attempted Reform of the SEBs

Through the 1980s and early 1990s, various efforts at SEB reform led by the central government, the World Bank, and independent researchers all suffered from either insufficient or weak implementation.[11] In 1991, the central government attempted to solve the problem of electricity supply to farmers. A committee recommended the establishment of a common minimum agricultural tariff, and a subsequent Chief Ministers' conference proposed that agricultural tariffs meet the modest target of 50 percent of the average cost of supply.[12] However, in the face of mobilized farmer vote banks, state governments took little action.

The World Bank provided loans to SEBs for financial restructuring, tariff adjustment, improved metering and collection, and other measures to increase distribution efficiency and revenue flow (World Bank, 1999a). In addition, World Bank support for NTPC was intended, at least in part, to promote good management practices within SEBs. By 1993, however, the World Bank had decided that SEBs had sunk into both a political and institutional quagmire and that institutional reform under the current ownership structure was a lost cause.

In 1991, an independent team of scholars published the DEFENDUS (DEvelopment-Focused, END-Use oriented, Service-directed) model, a unique Integrated Resource Planning approach that emphasized access, equity, and efficiency improvements.[13] Using this model, an analysis for the state of Karnataka showed that the requirements of electricity and installed capacity would only be about 40 per cent of what would be required in 2000, according to a conventional projection commissioned for the state. But administrators only seemed to have a perfunctory, academic interest in this approach, and in Integrated Resource Planning in general.[14] It was never seriously examined, despite several appeals to develop long-term electricity policy for the country.

By the beginning of the 1990s, there was broad consensus that the electricity sector was in dire straits and that the status quo was unsustainable, particularly in financial terms. If there was a moment to seriously consider re-regulation of the sector to reassert the independence of SEBs from their political masters, devise mechanisms of accountability, and cut through the Gordian knot of politically influential consumers pampered by subsidies, this was it. But the moment passed without any considered reflection about policy reform. With the growing consensus favoring a shift in macroeconomic policy, spurred by the balance of payments crisis, India was set to press the accelerator and motor into the next century. The electricity sector was at the forefront of the new liberalizing India.

A MANY-LAYERED REFORM PROCESS

The reforms themselves unfolded in four stages. In 1991, the central government invited private investment in generation. When this approach failed to address the root problems in the sector, a World Bank-supported reform effort in the state of Orissa, organized around unbundling and privatization in the sector, heralded a new stage in the reform process. This model was then followed by several other states. Finally, the central government reentered the debate by proposing a sweeping legislative reform package. (*See Box 4.2.*)

Attracting Private Investment: The IPP Debacle

In late 1991, the Ministry of Power swept away four decades of public monopoly in an act of great political significance. The new Independent Power Producer (IPP) policy was greeted with enthusiasm. However, little actual investment materialized, and a decade later, the IPP policy is broadly viewed as a flawed and halfhearted approach to reforms.

The Electricity Laws (Amendment) Act of 1991 allowed private entities to establish, operate, and maintain electricity generation plants as Independent Power Producers (IPPs) and to enter into long-term power purchase agreements with SEBs. Industry groups and urban middle class consumer groups welcomed the diminution of a public sector role and the entry of the private sector (Desai, 1999).

BOX 4.2	**CHRONOLOGY OF ELECTRICITY SECTOR REFORM IN INDIA**

- **1991** — Electricity Laws (Amendment) Act allows private sector participation in generation, with foreign investors allowed 100 percent ownership.

- **1992-97** — Eight projects given "fast-track" approval status and sovereign guarantees by the central government.

- **1995** — Orissa Electricity Reform Act established the Orissa Electricity Regulatory Commission and provided for unbundling of Orissa State Electricity Board.

- **1996** — World Bank support for Orissa Power Sector Restructuring Project approved.

- **1996** — Chief Ministers' Conference formulated a common minimum action plan for electricity.

- **1997** — World Bank Haryana Power Sector Restructuring Project approved, and Haryana state government passes the Haryana Electricity Reform Act.

- **1998** — Electricity Regulatory Commissions Ordinance Notification provides for establishment of a Central Electricity Regulatory Commission and state-level electricity regulatory commissions.

- **1999-2001** — Andhra Pradesh, Karnataka, and Uttar Pradesh proceed with preparation of Electricity Reform Acts. The World Bank prepares and approves projects supporting reform in each of these states.

- **2001** — Energy Conservation Bill passed by Parliament.

- **2000-2002** — Draft central government Electricity Bill prepared and introduced in Parliament.

Believing that private investors would be reluctant to come to India without generous incentives, the government acted with extravagance. IPPs were offered a guaranteed 16- percent return on equity, with bonuses for improved capacity utilization, a five-year tax holiday, and low equity requirements equivalent to 20 percent of project costs (Ahluwalia and Bhatiani, 2000). To further hasten implementation, the central government subsequently declared eight of the most promising projects "fast track" projects with expedited clearance procedures, and provided government counter-guarantees and escrow accounts against nonpayment of dues by SEBs. These incentives had the desired effect. By mid-1995, project developers and financiers had put forward 189 project offers totaling over U.S. $100 billion, which would have increased capacity by 75 gigawatts.

Believing that private investors would be reluctant to come to India without generous incentives, the government acted with extravagance.

Despite the initial "euphoric" reaction, as one senior bureaucrat put it, there were also early grumbles of discontent from various quarters, which steadily grew louder as the IPP policy failed to deliver (Pillai and Krishnamurthy, 1997). While supporting the policy, IPPs grew increasingly critical of bureaucratic delays and hurdles in implementation, and ever more concerned about recovery of dues from SEBs. In reaction, an Independent Power Producers Association of India (IPPAI) was established in 1995 to serve as a "neutral proactive forum."[15] IPPAI did create an important space for articulation of concerns about the reform process, although there was also a prevailing "negative opinion" within government ranks of IPPAI's perceived emphasis on winning special favors for IPPs.[16]

The central government was by no means unified on the IPP policy. The Ministry of Power was perceived as the primary promoter of the policy, with support from the Ministry of Finance. One widely held view was that although the IPP policy was "flawed," it had "been the most promising option at that time."[17] However, within each ministry there were stronger dissenting voices, with some at the Ministry of Finance who argued that concessions to IPPs might lead to net foreign exchange outflows rather than inflows. Moreover, the Ministry of Power's suspension of technical and environmental clearance for smaller projects aroused the ire of agencies responsible for those clearances.

Multilateral donors played a curious dual role in the IPP policy. While welcoming private electricity initiatives in principle (World Bank, 1991), the World Bank delivered a strong critique of the highest profile IPP, the Enron project, in a confidential memo to the Government of India. (*See Box 4.3.*) The memo stated that the project was "not economically viable, and thus could not be financed by the Bank," but urged the government to "explore ways to sustain the interest of the project sponsors" (Vergin, 1993). That the World Bank expressed its concerns about the project is laudable; that it did so only in a muted fashion is problematic. The IPP policy itself was widely viewed as faulty, since it threatened to further weaken the fiscal situation of states. Since the World Bank was actively supporting SEB reform at this time, it could well have been more public with its views. While there is no direct evidence on this point, Bank staff may have faced pressures to reconcile an IPP policy they viewed as flawed with the Bank's enthusiastic support for India's liberalization efforts. As a result, an important moment for critical reflection on the IPP policy was lost.[18]

The long-term impacts of the IPP policy were several and diverse, and are well illustrated by the high-profile case of the Enron project. First, key institutions responsible for long-term planning, and technical and economic clearance were weakened. Officials at well-functioning public agencies such as NTPC felt that the IPP policy created an uneven playing field in favor of foreign investors. Second, the reckless focus on capacity expansion excluded consideration of a more rational least-cost planning approach to electricity development. Finally, in its conception and implementation, the IPP policy offered opportunities for graft and malfeasance.

Projects were not typically selected through competitive bids, and power purchase agreements were kept secret even though they contained "take-or-pay" contracts involving public financial obligations for decades to come.[19] While no accusations have been conclusively proved, some high-profile projects have been caught in a swirl of accusations concerning human rights abuses, flawed environmental clearances, and corruption.[20]

Moreover, the IPP policy had a polarizing effect at multiple levels. Early support by urban middle class consumer groups and industry associations, who saw in the policy the promise of efficient power delivery, translated into anger toward public interest advocates who were seen as unnecessarily obstructionist (Desai, 1999). Within government ranks, those who saw the policy as the best option at the time were pitted against those who viewed the policy as flawed from the start. Thus, technically, economically, and politically, the policy created a hangover effect for future attempts at reform.

An Experiment with SEB Reform: The World Bank-Led Orissa Model

On a parallel track to the IPP process, the World Bank played a major role in arguing for fundamental reforms of SEBs, and in persuading a few states—led by Orissa—to initiate reforms. Having unsuccessfully tried in the 1980s to reform SEBs within the existing structure, World Bank efforts in the 1990s were directed at unbundling and privatizing SEBs. Hence, these reforms were considerably more far-reaching than the IPP policy.

Within India, there was broad agreement that the root causes of the problem were the technical, financial, and management problems of SEBs, but there was no agreement on the solution and on how to address the political thicket that SEB reform entailed. The World Bank stepped into this morass, armed with its new 1993 policy for lending to the electricity sector (World Bank, 1993a). At a workshop for Indian policymakers, the Bank highlighted the experience of ongoing reform experiments in the United States, United Kingdom, Argentina, and Chile. It offered to provide lending to support "...the boldest...most deserving state-level power sector reforms," but it would not finance or provide guarantees for electricity projects in states that did not undertake restructuring (World Bank, 1993b).

Of the few states that expressed interest in the World Bank's offer, the state of Orissa in eastern India was the first to proceed with a reform program. By the early 1990s, Orissa's electricity sector was in shambles. Transmission and distribution losses were estimated at 43 percent, only 17 percent of bills were collected, and the ratio of customers to staff was an astonishingly low 29:1 (Thillai Rajan, 2000, p. 660). However, the Bank selected Orissa mainly for political reasons. The Chief Minister of the state demonstrated strong political support for carrying through reforms.[21] Orissa also had a small electricity load in the agriculture sector and a weak farmer lobby (Thillai Rajan, 2000).[22] With low levels of political mobilization and a minor national profile, Orissa was "an experimental rat" for reforms.[23]

While local political support was undoubtedly necessary, the World Bank was the driving force for reform and the most consistent motivator of change.[24] For example, the Bank urged increases in tariffs to lay the groundwork for reforms.[25] World Bank staff candidly described their role as overcoming "natural resistance to change" within the state.[26] Reform consultants, NGOs, government officials, and the media eventually referred to electricity sector reforms in Orissa as the "World Bank model." These opinions were often not cast in a negative light, but as an appreciation of the Bank's proactive role in building momentum for change, and of the effort and commitment of particular staff members.[27]

The World Bank's "Orissa Power Sector Restructuring Project" required $997.2 million, and was partially funded by the then-Overseas Development Agency of the United Kingdom. Almost three fourths (74 percent) of the financing went to rehabilitation of distribution and transmission. A second component (23 percent) was allocated to demand side manage-

In October 1992, the Congress-led government of Maharashtra announced to the world that it had signed a memorandum of understanding with Dabhol Power Company (DPC), the Indian subsidiary of the U.S. based Enron Corporation, for a liquefied natural gas plant of 2,000 to 2,400 megawatt capacity, and to purchase electricity for 20 years. In what would later become a source of controversy, the deal was completed with alacrity and secrecy, despite the considerable size and financial obligations of the project, amounting to an expenditure of roughly $1.3 billion per year.

Despite strong reservations expressed by some state and central government bureaucrats, and by the World Bank, the project was cleared. Just as lending arrangements were being concluded, the newly elected state government, whose election platform in 1995 had stressed national self-reliance, canceled the contract and proposed to invite competitive bids. The international response was primarily negative, with concerns expressed about the viability of India's reform program and India's commitment to contractual obligations.

Yet, there were good grounds for concern about the project. Journalists and analysts found indications of complicity among officials to bend laws to accommodate Enron's demands and obtain the necessary clearances. Others predicted that the financial terms of the deal were highly unfavorable to the Maharashtra state electricity board, and that public funds were being jeopardized through the use of counter-guarantees. In addition, following charges of violence against opponents of the project, a Human Rights Watch investigation found that the state government had engaged in systematic suppression of freedom of expression and assembly, and that the Dabhol Power Company and Enron Corporation were complicit in these violations.

Despite this growing rumble of protest, within 2 months of the project being canceled, a new Power Purchase Agreement (PPA) was signed on the recommendation of a government committee with few changes to the original project. All clearances were subsequently awarded and counter-guarantees approved. Despite a pending public interest lawsuit challenging the final clearances that were given to the project and alleging fraud, the first phase of the project has been commissioned.

By 2001, the project had started to generate severe financial problems for Maharashtra. The SEB,

---

ment, with the remainder going to support the reform process (World Bank, 1996).

International consultants brought in by the World Bank and other donors played a considerable role in shaping reforms.[28] While consultants were hired for their technical knowledge, they frequently also had to assess the sociopolitical and institutional context for reforms. For example, consultants decided on a single-buyer system for Orissa, based on an assessment that the underlying technical, institutional, and commercial capabilities in the state were insufficient to support wholesale competition. In considering approaches to unbundling public utilities, they had to consider the need to minimize layoffs to avoid union opposition. Some national actors questioned the appropriateness and ability of international consultants playing these roles. One domestic public official said that consultants "sought to fit Orissa into their patterns," while another argued that their approach was like "applying principles of aviation to a jeep."[29] Some national consultants with considerable experience in the sector resented being placed in junior positions, although they were well-placed to educate international consultants on local conditions.[30] Since national and international consultants compete for contracts, these comments should not be uncritically accepted at face value. However, international

which had been profitable in 1998-1999, plunged into losses exceeding $300 million (excluding subsidies received from the state government) in 1999-2000. In order to honor its contract, the state had to buy power from the Dabhol plant at a cost twice that of the average production cost of electricity in the state.

Following a series of defaults on payment by the SEB, Dabhol invoked its financial guarantee from the state. When the Maharashtra government expressed its unwillingness to pay, the state's credit rating was downgraded. DPC subsequently invoked the counter-guarantee, by which time the SEB and the state government cleared their dues. Indeed, Enron officials mobilized senior U.S. government officials to raise the subject with the Indian government. DPC has since initiated arbitration proceedings in London, but the SEB has countered that the proper forum for settling all disputes with the company is the state regulatory agency, a dispute that has since moved to the Supreme Court. Most recently, with Enron Corporation itself in deep financial trouble, the troubled plant is up for sale to competing bidders.

Sources:

Enron Action Group. "The Enron Saga." Online at: http://altindia.net/enron (May 30, 2002).

Godbole Committee Report. 2001. *Report of the Energy Review Committee.* (April 10). Online at: http://altindia.net/enron/Home_files/doc/godbole/godbole-report.doc (May 30, 2002).

Human Rights Watch. 1999. *Enron Corporation: Corporate Complicity in Human Rights Violations.* New York: Human Rights Watch.

Mehta, Abhay. 1999. *Power Play: A Study of the Enron Project.* Hyderabad, India: Orient Longman.

Reddy, A.K.N. and A. D'Sa. 1995. "The Enron and Other Similar Deals vs. the New Energy Paradigm." *Economic and Political Weekly.* Vol. XXX (24) June 17: pp 1441-48.

Reuters News. 2001. "Moody's 'disappointed' in India's fiscal effort." (June 1) Online at: http://investor.cnet.com/investor/news/newsitem/0-9900-1028-6159822-0.html (May 30, 2002).

Sant, Girish, S. Dixit and S. Wagle. 1995. "The Enron Controversy: Techno-Economic Analysis and Policy Implications." Pune: Prayas. Online at: http://www.prayaspune.org/energy/eng_pub_sel.htm (May 30, 2002).

Weisman, Steven R. 2002. "Money, Energy Politics, and Enron's Costly Misadventure in Bombay." *New York Times.* (February 3).

---

consultants' own views suggest that there are downsides to reliance on expatriates. As one consultant put it, "subtleties...got past us."[31]

National actors—whether official or unofficial—did not substantially modify the consultants' proposals. The reform process was managed by a set of working committees, guided by a steering committee that reported to the Orissa Secretary of Power. The intent was to bring together government officials, SEB officials, and donor agencies. However, the reform committees, with limited experience with private ownership and competitive electricity markets, made few modifications to the consultants' proposals.

Consultations and a media campaign were intended to reach out to the broader public.[32] Critics of the consultation process charge that the goal was to "achieve consensus on a model rather than to evolve a model through a consensual process."[33] Interviews support this view. Participants saw the role of consultations as explaining changes and "reducing tension."[34] NGOs reported that their concerns—including the impact on access for electricity to the poor—did not result in any changes to the approach.[35] Indeed, the process appeared designed to usher reforms through rapidly, based on a political judgment that a long process would allow vested interests time to mobilize opposition to reforms.

The Content of Reforms

Reforms in Orissa, following the Bank's approach being implemented in much of the world, consisted of:

- unbundling generation, transmission, and distribution;
- allowing for private participation in generation and transmission utilities;
- privatizing existing thermal generation and distribution utilities;
- establishing an autonomous regulatory agency; and
- reforming tariffs at the bulk electricity, transmission, and retail levels.[36]

The lynchpin of the reform process was the passing of the Orissa Electricity Reform Act in 1995, which provided for the establishment of an independent regulatory commission and the divestment of equity in generation and distribution to the private sector.

NGOs reported that their concerns—including the impact on access for electricity to the poor— did not result in any changes to the approach.

Public officials and Indian consultants suggest that the reforms were single-mindedly focused on financial issues and on privatizing the sector. According to one official, international donors were obsessed with removing subsidies and increasing tariffs.[37] Another characterized the donor approach as "privatization must be done; let's do it somehow."[38] A representative of a donor agency confirmed this perspective when he described the Orissa reforms as "basically a bankruptcy workout."[39] International consultants emphasized that they received instructions to promote rapid privatization, and to "create a process that was irreversible."[40,41] Donor agencies saw financial issues at the heart of the restructuring and enhanced private participation in the sector as the best solution. It was anticipated that private finance would develop new generation capacity and enhance availability of existing capacity. Private participation in distribution was expected to improve service quality and increase financial performance. Donor agencies were not alone in this view. Some senior national and state officials held the same position. Others reluctantly agreed, only because they felt that all other options, notably continued public ownership, had been exhausted.[42]

Yet, attracting investors for privatization in Orissa proved to be a difficult task. To make the distribution sector more attractive, 75 percent of the shared financial liabilities were transferred to the publicly held transmission sector.[43] To make generation more attractive, generation companies were allowed to increase the price they charged to the public transmission company, but the transmission company was not allowed to pass on higher prices to distribution companies. As a result, the only public component, the transmission company, built up enormous liabilities that undermined its long-term viability. Ultimately, privatization was carried out, but there was limited interest and few bids.[44]

The results have not been positive. Since privatization, the new owners have brought neither new funds nor discernible management skills to the newly established companies.[45] Revenues from privatization were not plowed back into the sector, but absorbed into the government budget for other purposes.[46] The public has faced substantial tariff increases but seen few benefits in service, which has led to growing political discontent with the reform process and a call to bring back the publicly owned system. The private operator of one distribution zone, which also operates one generation unit, believes that the government has neither ceded management control nor paid its own bills.[47] As a result, this company has taken steps to withdraw from the sector in Orissa. Consequently, the Government of Orissa established a high-level committee to reconsider the reforms. The committee found that the new distribution companies had failed to bring in significant additional financing and that reductions in transmission and distribution losses had been minimal.[48] Despite these problems, the fact that

Orissa had embarked on and been through several stages of a reform process, including privatization, provided a powerful demonstration effect within India. Other states soon lined up to follow Orissa's lead.

Scaling up the Model

By 1998, Orissa had managed to demonstrate that it could privatize its distribution business, and the more problematic aspects of the Orissa experiment had not yet materialized. Growing disenchantment with the IPP policy left states with few alternatives other than reform of SEBs to address an electricity sector crisis that showed no signs of abating. Moreover, as economic liberalization grew more palatable, opposition to privatization faded. Even states with avowedly communist governments competed to invite private investors (Echeverri-Gent, 2000). Finally, the World Bank continued to stand ready to support states that wished to embark on a reform program. As a result, since 1995, several large and politically significant states have concluded (or are in an advanced stage of negotiating) loan agreements with the World Bank to reform their electric power sectors.

These states have followed the basic parameters of the Orissa model, in many cases guided by the same consultants, but there have also been some significant differences. First, in subsequent efforts, electricity reforms have been part of the broader framework—articulated in the World Bank's Country Assistance Strategy for India—of state-level financial restructuring. This approach is relatively new for the World Bank, since it involves providing a broad macroeconomic restructuring loan at the state level rather than to a national government. Second, most of the new World Bank loans are structured as "Adaptable Program Loans" (APLs) that release small amounts of funds over many years, with each tranche dependent on the fulfillment of conditions. Compared to a single large loan, this approach enables the World Bank to provide a down payment on future support, to signal seriousness of intent to investors, and to provide the World Bank flexibility in adapting to future conditions (World Bank, 1997b).[49]

Finally, in response to difficulties faced by private distributors in Orissa, subsequent efforts have sought to mitigate risks that tariffs will not be raised, payments will not be collected, or thefts will not be reduced.[50]

The World Bank has not been the only donor agency active in the sector in India. The U.K.'s Department for International Development (DFID), Canadian International Development Agency (CIDA), U.S. Agency for International Development (USAID), and Japanese aid agencies have also provided funding for elements of the reform. Of these, DFID has provided considerable funds for technical assistance with the reform program (World Bank, 1999b). Much of DFID's contribution has been in the form of a grant rather than a loan. According to one World Bank observer, DFID's grant support for basic technical work was critical to implementation of reforms.[51]

It is important to note that not all states have decided to follow Orissa. A few states, including Gujarat, Madhya Pradesh, and Tamil Nadu, have decided to focus on commercialization of their SEBs rather than going down the road toward privatization. In some cases, they are receiving support from the Asian Development Bank. While it is too early to compare experiences across states, in the future these varied approaches will provide valuable material for a comparative assessment.

The Central Government Follows the Lead of the States

With many states following the Orissa approach, the central government took steps to provide a legislative framework for state-level reforms. In 1998, the Ministry of Power championed an Electricity Regulatory Commission Act, creating a central regulatory agency and providing an umbrella framework for each state to establish its own agency.[52] This act marked the first formal sign of recognition by the central government of the significance of Orissa's reform efforts, and was a late effort to provide a template for state-level reforms.

In 2000, the Ministry of Power initiated the drafting of a comprehensive Electricity Bill to replace all existing legislation in the sector. This bill is the most dramatic initiative taken to date by the central government to exercise some leadership over the direction of the sector. In contrast to the state reforms, preparation of this bill has been a domestic effort, initiated and led by the Ministry of Power. The World Bank has limited itself to comments on drafts. The bill requires states to unbundle their SEBs, establish independent regulatory commissions, facilitate open access to transmission (wholesale competition), develop a spot market for electricity, and meter all electricity supply (Suri, 2000). Although the Ministry of Power now does support privatization, the bill does not explicitly require privatization, but gives the states some flexibility on how to organize ownership of an unbundled sector.

Plans to introduce the bill in Parliament, originally intended for 2000, were postponed after the sudden demise of then-Minister of Power Kumaramangalam. In the interim, the debate has been shaken by the tumultuous experience with post-reform competitive electricity markets in California and the meltdown of the Enron Corporation. In particular, ambitious market frameworks such as spot markets for electricity have now been placed on the back burner (*Economic Times*, 2001).

The central government has sought to promote fiscal responsibility. For example, a central government-convened expert group recommended in mid-2001 that SEBs take responsibility for past dues, and that incentives were needed to support this effort.[53] They also argued that failure to service future obligations should meet with heavy censure.

Central government direction has also led a broad trend away from acceptance of electricity provision as a purely commercial enterprise, and more willingness to reinsert social and economic development goals within a broad framework of fiscal accountability.[54] For example, a ministerial committee has promoted a concerted dialogue on rural electrification in the context of the electricity bill. This committee is likely to embrace a system of decentralized licenses managed by state electricity regulatory commissions for rural electricity provision, and introduction of a system of subsidy auctions—inspired by experiences in Argentina and Chile—for those willing to undertake rural electrification.

In addition, evidence of a more proactive approach to environmental considerations as they relate to the fiscal and other goals of reform have begun to surface. For example, the Ministry of Non-Conventional Energy Sources has proposed that a preferential tariff be introduced for wind energy projects, and that the Electricity Bill mandate that a minimum of 10 percent of electricity generation be obtained from renewable sources (*Bulletin on Energy Efficiency*, 2002a). Few developing countries have pursued such an approach, although China is among this small group. *(See Box 4.4.)* In addition, an Energy Conservation Bill was passed by Parliament in August 2001. It calls for the establishment of institutional and legal structures to implement energy efficiency, relying on both regulatory enforcement and market inducements (*Bulletin on Energy Efficiency*, 2002b).

With regard to the broader reform agenda, the debate appears to have shifted from the far-reaching goals of instituting complex spot markets to using the Electricity Bill to meet more pressing demands. These include the long-standing objectives of metering all consumers, increasing tariffs and removing cross subsidies, and reducing transmission and distribution losses. Since implementation of this agenda will require considerable funds, the course of actual reforms will be dictated by the availability of financing. In this context, the World Bank's policy of making funding conditional on private participation in the sector takes on renewed significance. Only states that signal willingness to privatize will have access to external funds.

BOX 4.4 — CHINA'S EXPERIMENT WITH A RENEWABLE PORTFOLIO STANDARD

In mid-2000, the Chinese government announced that it was considering adopting a renewable portfolio standard (or RPS) as a part of its Tenth Five-Year Plan for National Economic and Social Development covering the years 2001-05. Under the proposed RPS system, about 5.5 percent of each province's and autonomous region's electricity would be required to come from renewable energy sources. This plan, it was hoped, would provide incentives to replace coal-fired power plants with renewable energy sources, leading to a reduction in air pollution. Because many of the sources of China's renewable energy—wind farms, small hydroelectric dams—are in the relatively impoverished west, those designing the RPS system expected that provinces in the relatively wealthy east would have to buy electricity from western producers to meet the 5.5 percent mandate, thereby realizing a politically desirable transfer of wealth from east to west.

When the official Tenth Five-Year Plan for Economic and Social Development was announced, it contained a single sentence that urged the government to "implement favorable pricing for bringing new and renewable energy sources onto the electricity grid, and to support the prompt development of the Mandated Market Share (RPS)." Although it is important not to read too much into one sentence, it is nonetheless significant that this explicit endorsement of an RPS came from the National People's Congress, the highest organ of state power in China.

There remain significant obstacles to China's implementation of an RPS in the near future. Primary among them is a need to pass a law implementing a RPS. In addition, implementation will require creating a unified national power grid (as yet unrealized); convincing provincial authorities that run the grids to honor the terms of private energy contracts; and establishing a competitive national energy market. None of these reforms will be easy. Nonetheless, observers both inside and outside China have voiced cautious optimism that the government's flirtation with an RPS may yet lead to a cleaner, greener electricity sector for China.

Sources:

China Daily. 2001. "Premier's Report on Outline of New 5-Year Plan at the Fourth Session of the Ninth National People's Congress on March 5, 2001." Online at: http://www1.chinadaily.com.cn/highlights/docs/2001-04-30/3550.html (March 5, 2001).

China Internet Information Center. "Look in the Next Five Year Plan: Blueprints." Online at: http://www.china.org.cn/e-15/15-3-b/15-3-b-31.htm (May 21, 2002).

Hamrin, Jan. 2002. Centre for Resource Solutions. Personal communication. (May 23).

Martinot, Eric. 2001. "Renewable Energy Market Development in China." Presentation made at Woodrow Wilson Center, Washington, D.C., July 19.

United States Embassy in China. 2001. "Energy Goals for the 10th 5-Year Plan." *Beijing Environment, Science and Technology Update.* September 7. Online at: http://www.usembassy-china.org.cn/english/sandt/estnews090701.htm (May 21, 2002).

In sum, central government efforts to steer reforms do provide an opportunity to step back from the Orissa model driven by narrow financial considerations and think through the broader objectives of reform. However, it is not clear how these efforts will mesh with World Bank-funded state reforms, which so far have been focused on financial restructuring.

THE ROLE OF PUBLIC BENEFITS IN THE REFORM PROCESS

It is far too early to conclude whether social and environmental conditions on the ground have improved as a result of the reforms. But a close look at the process provides insights into whether and how public benefits were factored into the decisionmaking process by the major actors involved.

Social Issues through a Fiscal Lens

To the extent social issues have been raised in the reform context, they tend to be viewed primarily through the lens of better fiscal management. The World Bank, in particular, suggests that reforms in the electricity sector would free state funding for "higher priority use in the social sectors" (World Bank 1999b, p. 27). Thus, the framing of the electricity sector largely excludes explicit consideration of its social dimensions, a break with the previous rationale for state involvement in the sector. Where social considerations are explicitly addressed, the reform loans do not build in measures to ensure they are achieved.

For example, the World Bank emphasizes the importance of defending concessional rates for low-income groups in the face of price increases (World Bank, 1999b). Yet, it is not clear how the continuation of lifeline rates, which will continue to place burdens on the state exchequer, can be reconciled with a desire to free funds for allocation to other priority social sectors. The magnitude of the financing shortfall is well illustrated by Ahluwalia (2000), who computes that about 50 percent of all households (81 million households) are unable to afford commercial rates for electricity.[55] Hence, even though the current burden on the state budget comes largely from a debt service obligation, even if these were to be minimized, social spending in the sector could easily consume much of the savings. If these households are to be provided electricity at affordable rates, there should be no illusions about the continued need for public funds even in a privatized and restructured sector.

On the important question of increasing access to electricity services, World Bank loan documents note that the commercial orientation introduced by the reforms will lead to more modest targets. At the same time, they argue that the enhanced efficiency of the resultant institutions will lead to more effective implementation on the ground, more than compensating for the lower targets (World Bank, 1996; World Bank, 1999b). Yet, since the private sector is unlikely to invest in connecting low-income, and typically loss-making customers, it is unlikely that even modest targets will be met without a financial incentive. Hence, a strong case can be made that reforms—whether the sector is under public or private ownership—should be accompanied by intentional efforts to provide incentives for increasing access. Of the various actors in the reform arena, only the central government has shown any interest in exploring the potential for such schemes. As yet, however, no concrete measures have been taken to address the problem of limited access to electricity.

Finally, there is one hopeful outcome from the privatization experience in Orissa. Privatization has allowed decentralization of distribution responsibilities with an attendant improvement in performance. For example, the local Xavier Institute of Management (in collaboration with the Bombay Suburban Electricity Company) has established village collectives to manage and organize bill collection tasks in a few pilot rural areas. The initial experience suggests that rural residents respond very positively to control over electricity management at the village level. For example, newly formed village committees achieved a 100 percent increase in bill collections over a six-month period.[56] Certainly, this approach needs to be subject to greater scrutiny to ensure that decentralization does not transfer power into the hands of local elites. Nonetheless, this limited experience does suggest that aside from the debated benefits of privatization, there are potential collateral benefits arising from the greater scope for decentralized forms of organization in the sector following a loosening of state control.

A Restricted View of Environmental Costs and Benefits

The World Bank is the most explicit of the various actors on the need to address environmental concerns. However, discussion of the environmental implications of reform is driven by the World Bank's internal "safeguard" policies, which are designed to ensure that negative effects of investment projects are guarded against and mitigated. Within this framework, environmental impacts refer rather narrowly to the direct environmental impact of loan funds spent on physical infrastructure, such as resettlement due to power plant construction, land acquisition for transmission lines, and the like. This narrow interpretation fails to account for environmental impacts of the broader regulatory reform put in place through the reform process. Consequently, the Bank's interpretation of its environmental guidelines hew to a rather narrow do-no-harm approach, rather than looking for environmental gains through reforms.

The World Bank did conduct a substantial study on environmental issues in the electricity sector (World Bank, 1998).[57] The study notes that the sector is on the verge of massive changes, but it explicitly does not address the environmental impacts of the institutional and managerial dimensions of reform—such as unbundling or tariff liberalization—or the implications of changes in ownership from public to private. Instead, the focus is on the environmental impacts of implied changes in technology and in the price of electricity. Other than encouraging attention to demand side management (*see below*), there is little evidence of the impact of the study on the design of state-level reform packages and associated World Bank loans.

Sources within the World Bank place responsibility for the limited scope of the study with the Ministry of Power. When the study was in progress in the mid-1990s, the Ministry was not convinced of the value of institutional reform. With this mindset, they were concerned that such a study could lead the Bank to impose environmental conditions on reforms, and that the study would contribute to a consensus favoring one particular route forward for state-level reforms, pre-empting a broader debate.[58] This limited the scope of the study. While the environmental issues study does provide useful information on the relative costs and benefits of specific technological measures, the inattention to institutional changes was an opportunity lost.[59]

At the state level, the only concrete attempts to implement an environmental component to the reforms involved promoting demand side management (DSM). In Orissa, the World Bank, which allotted 13 percent of the reform loan to DSM efforts, led this effort. This enthusiasm was driven in part by the demonstrably large potential for DSM in India. It was also a political reaction to fierce criticism of the Bank for its lending program in India, particularly for the controversial Narmada valley dam projects. However, there was widespread skepticism about DSM among other donor agencies, international reform consultants, and state officials, who cynically viewed DSM as a measure to satisfy internal Bank politics and procedures—"a box to be checked."[60]

For two reasons, the results in Orissa were not encouraging. First, the technical scope for DSM in Orissa was limited. Orissa had surplus electricity at the time of reforms, and there was no incentive for the utility to reduce consumption by paying customers. In addition, Orissa had a small agricultural sector. In other states, the agricultural sector is a prime candidate for DSM, since it is a loss-making sector for the utility. Second, DSM staff complained that they received little political support from the World Bank, and this view of DSM as an "embellishment" percolated through to consultants and public officials. As a result, despite the allocation of substantial funds, even the opportunities that were available were not taken.[61]

DSM has remained on the agenda for other states, where it is a more timely idea. Moreover, support for the idea has deepened and broadened within the World Bank and within India. Implemented correctly, DSM could ameliorate supply shortfalls and build a political constituency for reforms—particularly in rural areas—by bringing demonstrable benefits early in the reform process. However, the lesson of the Orissa experience is that realizing both technical and political benefits requires more political support and attention to DSM as an integral part of reform efforts.

Finally, as discussed earlier, the national government has attempted to promote renewable energy technologies and energy efficiencies through various legislative instruments. These efforts represent an encouraging attention to environmental concerns, but have not yet led to any concrete gains.

Innovations in Governance: The Emergence of an Independent Regulatory Culture?

Since past problems in the electricity sector are directly associated with the effective capture of electricity sector institutions by vested interests, regulatory commissions are a lynchpin in a new model aimed at independent operation. The first regulatory commission set up for the electricity sector, the Orissa Electricity Regulatory Commission (OERC), has set impressive standards for transparency in India. So far, its performance with respect to access to information and consultation has been strong. Notably, the OERC has set up a comprehensive web site to disseminate information. On several issues, the Commission has held open hearings, where labor and consumer groups have spoken.[62]

With regard to independent operation, the central issue for state regulatory agencies has been their control over tariffs (Balakrishna, 2000; Indiapoweronline.com, 2000b). In some states, notably Orissa and Maharashtra, regulators have been reluctant to allow tariff increases without evidence of reduced losses. Regulatory decisions on tariffs have not gone unchallenged. In Orissa, the World Bank explicitly urged the OERC to approve tariff increases to "provide comfort" to investors just before privatization, a request that they rejected.[63] In Madhya Pradesh, the regulatory agency refused to allow tariff hikes, a decision that was challenged by the state government. In Andhra Pradesh, in contrast, tariff increases were strongly opposed by the public and by opposition parties (Indiapoweronline.com, 2000a).

State regulatory commissions exhibit a remarkable diversity of operation, particularly in the vigor with which they have defended their independence. Some State Electricity Regulatory Commissions (SERCs) are termed "mere extensions of government," at least in their regulatory culture, because they do not hold open hearings and tend not to pay attention to stakeholder comment or complaints.[64] In other cases, there is an active interest in seeking technical assistance and informal consultation from analysts and consumer groups, resulting in bold initiatives that annoy donors and state governments because they are seen as beyond the regulators' mandate.[65] In one instance, a consumer advocacy group has even provided regulators with analysis of utility performance.[66] Curiously, most regulators have come from bureaucracies with no great tradition of independence or public participation and consultation. State regulatory commissions have included as members former civil servants, judges, and former central or state electricity agency members with technical expertise. Yet, in some cases, as with the Orissa regulator, they have enthusiastically assumed the role of principled public oversight.

At the same time, critics have pointed out that the provisions requiring transparency and public consultation that guide regulatory functioning are by no means sabotage-proof (Dixit, Sant, and Wagle, 1998). The pressures for political accommodation remain as strong as before, as both regulators and government officials unofficially acknowledge. As one official put it, "There is not only one God in the Indian pantheon. Any regulator who does not talk to the government is living in a fool's paradise."[67] In this context, principles of good governance are diluted by granting the regulators discretionary powers, which allow them to circumvent application of these principles in a variety of ways.

Most significant is the zeal with which members of the public, including consumer advocates, environmentalists, the media, and even casual observers, have greeted the new institutions. There is keen interest among members of the public to "democratize" the commissions at an early stage. At the same time, few civil society groups are equipped to deal with the complex technical character of the sector, which can limit the degree of engagement with regulators.

While public participation is a necessary component, it cannot substitute entirely for public policy direction. Indeed, governments must give regulatory agencies appropriate guidelines on how to make the difficult political tradeoffs between economic, social, and environmental implications of their decisions. Unfortunately, state governments perceive the reform process as an opportunity to rid themselves entirely of what has become a burdensome sector, leaving an absence of responsibility for longer-term and broader issues raised by electricity sector development. Drawing from the experience of the first regulatory agency in Orissa, regulators are setting a precedent of ignoring these tradeoffs by limiting themselves to economic decisions, and in particular to a tariff-setting role, to the exclusion of the broader landscape of electricity development in the state (Sankar and Ramachandra, 2000). Part of the problem lies in the enabling legal framework, which does not empower regulators to address economic regulation and its economic and social effects in an integrated fashion. Despite this limitation, electricity regulators do occasionally embed environmental concerns in regulatory decisions.[68]

Moreover, the training that regulatory commissions receive on regulatory practice does not focus on the linkages between economic regulation and environmental outcomes. This training is often dispensed by international consultants with narrowly defined terms of reference, whose ranks are staffed by regulatory economists with neither the expertise nor the inclination to explore broader issues of public benefits in the sector. Since the past ills of the sector were perceived as the result of mixing social development with the business of providing electricity, the message typically delivered to the regulator is "it is not your role to solve social problems."[69] Yet, at the moment, there is no other body in a position to do so. Early attention to these issues is necessary because it will be hard to graft attention to public benefits onto the mandate and expertise of regulatory commissions at a later date. The initial period not only develops skills, but also sets priorities and shapes institutional cultures. The lack of attention to a long-term vision could ultimately limit the full potential of regulatory commissions as a progressive force in the sector.

CONCLUSION

Electricity sector policy in India has been locked into adverse arrangements at least twice in its history. The first was when agricultural consumption was demetered and extensive subsidies were offered. The second was when SEBs signed IPP contracts with major fiscal implications. A third set of circumstances, with the potential for equally powerful forms of institutional rigidities, are in the making with the reproduction of the Orissa model on a national scale. These circumstances may yield favorable institutions, like democratic and transparent regulation, but may also result in unfavorable ones, such as locking out integrated resource planning or scaling back programs to expand services to rural areas.

The World Bank has played a central role in moving the sector to the threshold of a new organizational form. The Bank forcefully argued that the sector had reached the end of its current road, and backed up this assertion by conditioning funds on bold reforms.

The Bank's success has rested only in limited part on the brute force of conditionality, and rather more on skill in building what appear to be genuine constituencies for reform among bureaucrats and politicians. Nonetheless, it is problematic that the Bank's dexterity led to the adoption, without broad public debate, of what appears to be a single dominant approach to transformation of a critical sector. That a few other states have adopted a different route based on reform, but no change in public ownership, will provide an interesting basis for comparison in a few years. By the time reform was served up to the nation in the form of the Electricity Bill, many of the key decisions had been made. A broader debate about the ultimate goals of policy change and the best means to achieve these goals could not only broaden the range of ideas, but also mobilize new actors to play a role in the regulatory process and build a constituency for reform. While the World Bank and its supporters have argued that opening a debate would condemn the sector to paralysis, the back-door approach, particularly in the early days of reform, limited participation in the debate to a few technical and financial experts. More recently, there are welcome signs that state-level reforms are subject to an open and more broad-based debate.

This review of the reform process suggests that there was little explicit attention paid to either the social or environmental dimensions of a public benefits agenda. While social issues received lip service, few measures were put in place to ensure that these objectives would be realized. With the exception of a genuine effort at demand side management at the state level by the World Bank, discussion of innovative financing schemes for ensuring rural access by the central government, and some discussion of incentives for renewable energy, there has been little explicit attention to environmental

outcomes. While reforms may yet indirectly lead to both social and environmental gains through the construction of a better functioning sector, there has been little attempt by any of the reformers to ensure this outcome. It is by no means clear that a long-term social and environmental vision can be subsequently woven into the fabric of reforms. Nor is it fully clear that social and environmental benefits are always coterminous with the techno-managerial vision of the sector based on privatization and a measure of competition. The history of agricultural subsidies and the IPP debacle should teach us how expedient choices in the present constrain our collective future.

Looking forward, considerable hope rests on the new autonomous governance structure of the regulatory commissions. Still, even that expectation is only tenuously sustained by the experience in some states, and less so in others. With regard to actively shaping a visionary future, independent regulation so far does offer many opportunities to promote public benefits. While enabling legislation provides some room for interpretation, regulators seem inclined to define their job narrowly, an inclination that is reinforced by the international consultants who train them. A conservative and narrow regulatory culture could be a particularly significant force for institutional lock-in that will shape the future development of the sector.

It is late, but perhaps not too late to have an informed public debate about the future of the sector. Such a debate should actively consider increased access to electricity, social pricing, and the promotion of sustainable energy futures as concerns to be integrated into reforms. Such debates could favor a decision to pursue short-term financial motivations first, as those who have led reforms thus far suggest. But it is also possible that broad dialogue will both enhance scrutiny over and the effectiveness of existing reforms, and suggest ways to achieve both short-term financial health and longer-term social goals. Either way, without explicit attempts to bring diverse groups into the debate, in a democratic polity the political sustainability of policy reform will always hang in the balance.

1. This chapter is a substantially revised and updated version of Navroz K. Dubash and Sudhir Chella Rajan, "Power Politics: The Process of Power Sector Reform in India," *Economic and Political Weekly*. XXXVI (35): 3367-3390.

2. There were clear warnings from the donor community that only about one fifth of required finance for developing countries' projected electricity needs would be available from multilateral sources (Churchill and Saunders, 1989).

3. A small number of private companies continued operation, particularly in large cities, largely buying electricity from SEBs.

4. SEBs are expected to operate on a commercial basis and earn at least a 3-percent return on their net fixed assets.

5. Interview with a former member of the Planning Commission, July 18, 2000. All interviews for this chapter were conducted on a not-for-attribution basis. Consequently, interviewees are identified only by their institutional affiliation.

6. The World Bank was supportive of this move, and directed more than half of its total allocation of $7 billion in sector funding between 1970 and 1991 to NTPC (World Bank, 1999a).

7. This record has been tarnished recently by the reports of human rights abuses at Singrauli in Madhya Pradesh (World Bank, 1997a).

8. This figure, drawn from various Central Electricity Authority surveys, must be treated with some caution. That distribution lines had reached most hamlets did not mean that all households were necessarily able to access and use electricity.

9. Interview with former Andhra Pradesh state official, July 20, 2000.

10. Sant and Dixit (1996) suggest that the benefits flow largely to landed farmers who can afford mechanized irrigation, and who use irrigation to grow high-value cash crops. Landless populations do not benefit from this policy (Verma, 1999), unless it is indirectly through greater employment opportunities. Moreover, cheap electricity encourages profligate use of groundwater, and large farmers are better able to mobilize capital to chase the water table than are small farmers (Dubash, 2002).

11. For example, see Government of India (1980) and Planning Commission of India (1994).

12. Reported in Government of India (1999).

13. Reddy et al. (1991). DEFENDUS modeled its framework on energy services by examining supply expansion as well as efficiency improvements, and allowed for environmental costs to be internalized. A committee for the Long-Range Planning of Power Projects (LRPPP) set up by the government of Karnataka projected that the state would require a six-fold increase in electricity supplies by the year 2000—from the 1986 consumption of 7.5 terawatt hours to 47.5 terawatt hours, and from the 1986 installed capacity of 2,500 megawatts to 9,400 megawatts. With end-use efficiency improvements, the DEFENDUS scenario proposed 17.9 terawatt hours of electricity and an installed capacity of 4,000 megawatts by 2000, together amounting to roughly one third of the cost of the original scenario.

14. Within the U.S. context, IRP has a rather specific meaning applicable to traditional (vertically integrated) utilities, which are required to submit plans to regulators for integrating demand side as well as generation options in their tariff submissions. We use the term here more broadly to refer to any attempt to identify, analyze, and acquire cost-effective resources, which would lower the long-term cost of energy services. In this definition, long-term resource planning (taking into account supply side and demand side efficiencies) would be conceivable even in an unbundled situation as long as a regulator could develop and implement incentives structures to promote more cost-efficient resource use.

15. See http://ippai.org.

16. Interview with government bureaucrat, July 20, 2000.

17. Interview with power sector official, July 13, 2000.

18. It is important to bear in mind, however, that sections of civil society were very active during this time in formulating their own responses to IPP policy. In various newspaper and magazine articles and other public forums, journalists, former bureaucrats, academics, and environmentalists criticized specific projects as well as the overall framework. One group of critics formed a "National Working Group on Power," and organized workshops and campaigns against IPP policy. Public interest litigation was filed on behalf of citizens against the government as well as specific IPPs on grounds of corruption, environmental damage, and constitutional violation.

19. The World Bank held a workshop on competitive bidding at Hyderabad in 1994, (personal communication, World Bank staff, February 2002). Nonetheless,

by then many of the largest power purchase agreements were negotiated in secret and without competitive bidding (Reddy and D'Sa, 1995).

20. For example, in the case of the Mangalore Power Corporation, where Cogentrix Corporation was the developer, public interest litigation was filed by a consumer activist in the Karnataka High Court alleging offshore payments by Cogentrix's partners through a subsidiary in the British Virgin Islands. The company has since withdrawn from the Karnataka project, citing delays in gaining government approvals and in resolving the litigation (Fernandes and Saldanha, 2000).

21. The initial reason for support was the World Bank hint that funding for a favored hydroelectric project would be more forthcoming if the state undertook broad reforms. While this tactic was instrumental in initially getting the Chief Minister's attention, several interviews with senior state officials (July 20, 2000) indicate that he very quickly developed a deep personal belief in the need for fundamental reforms in the sector. Thillai Rajan (2000) confirms this account.

22. Agriculture accounted for 6 percent of load in Orissa versus around 40 percent in many other states (Thillai Rajan, 2000).

23. Interview with power sector official, July 14, 2000.

24. Interview with Orissa state official, July 20, 2000, and interview with former national power sector official, July 18, 2000.

25. Interview with former national power sector official, July 18, 2000.

26. Interview with World Bank staff, July 13, 2000.

27. Interview with public power sector officials, July 18, July 20, July 25, 2000.

28. Interview with reform consultant, September 13, 2000.

29. Interview with Orissa public sector officials, July 25, July 26, 2000.

30. Interview with Indian consultant, July 15, 2000, and with academic, July 26, 2000.

31. Interview with international reform consultant, September 13, 2000.

32. World Bank (1996), and interview with Orissa public official, July 25, 2000.

33. Interview with NGO staff, July 22, 2000.

34. Interview with academic, July 26, 2000.

35. Interview with academic, July 26, 2000.

36. See World Bank (1996).

37. Interview with consultant, July 18, 2000.

38. Interview with Orissa power sector official, July 15, 2000.

39. Interview with donor agency staff, December 7, 2000.

40. Interview with international consultants, September 13, 2000.

41. Indeed, the Bank adopted a reform mantra, "Failure is not an Option" to emphasize "the importance of relentless pursuit of reform implementation at times of difficulties." World Bank (1996, Annex 5.3 p.4).

42. Interview with Orissa public official, July 20, 2000, and July 25, 2000, and interview with former national power sector official, July 27, 2000.

43. For details see Mahalingam (2000).

44. One company, Bombay Suburban Electricity Supply purchased three of the four distribution zones, and sought to purchase the fourth, but was turned down in order to introduce some competition (Mahalingam 2000, p. 96).

45. Interview with former Orissa power sector official, July 25, 2000.

46. According to one report, only 3 percent of the privatization revenues from the sale of the Orissa Power Generation Corporation were re-invested in the sector (Indiapoweronline.com, 2001a).

47. Personal communication with international reform consultant to Orissa, September 18, 2001.

48. See Pragativadi.com (2002).

49. For example, the Andhra Pradesh Adaptable Program Loan was structured around five sets of conditions: (1) pass a reform bill and reform tariff setting; (2) notify the bill, establish a regulatory commission, and unbundle the SEB; (3) partially privatize distribution; (4) further privatize distribution and list shares of the generation company on the stock market; and (5) privatize distribution completely and list shares of the transmission company (World Bank, 1999c).

50. For example, in the state of Karnataka, one proposed mechanism is the introduction of a "distribution margin" that guarantees income to the distribution company during a transition phase. This approach has been criticized as unduly insulating the private investor from risks that are within their ability to manage, and potentially limiting the authority of the regulator (Menon, 2002).

51. Interview with World Bank staff, July 6, 2000.

52. Under the Act, each state had the choice of establishing a commission on the basis of the central govern-

ment Act or through state level legislation, as Orissa had done.

53. Specifically, the committee proposed that SEBs be allowed to issue bonds in favor of creditors, the incentive being a waiver of 50 percent of the interest on past dues of SEBs. While this broad approach has been welcomed, whether it adequately recognizes the challenge to states to meet future obligations has been questioned (Ahluwalia, 2001).

54. Interview with power ministry official, March 8, 2000.

55. Most of these 81 million households currently do not have access to electricity. If the considerable challenge of providing them access to electricity is met, and these households were asked to meet half the average cost of supply, the remaining subsidy burden on the treasury would be about $1.4 billion. This is approximately the amount now spent on electricity subsidies, an amount which clearly does not reach the poorest and most needy. However, as this calculation suggests, the issue is not whether subsidies will be needed, but how they should be best targeted to reach the poorest.

56. Interview with staff from Xavier Institute of Management, July 26, 2000.

57. The study develops a methodology that is applied for two states, Andhra Pradesh and Bihar.

58. Interview with World Bank staff, July 6, 2000.

59. Indeed, sources within the government do suggest that World Bank studies—such as an early study on long-term issues in the sector, or an ongoing study on farmer uses of electricity—are influential and useful in internal debates (Interview with government official, July 14, 2000).

60. Interview with international consultant, September 13, 2000, and interview with donor agency staff, July 17, 2000.

61. Interview with reform consultant, September 16, 2000.

62. Interview with representative of consumers group, July 25, 2000.

63. Interview with public official, July 20, 2000, and with consultant, September 13, 2000.

64. Interviews on July 27, 2000 with consumer advocate and consultant.

65. Donor interviews, July 15-17, 2000.

66. In interviews, Prayas, a nongovernmental organization in Pune that has focused on sustainable energy issues, was referred to by officials and regulators as a credible NGO actor in the sector. Interview with government official, July 14, 2000 and interview with regulatory official, July 20, 2000.

67. Interview with former public official, July 20, 2000.

68. For example, regulators provide incentives for improved efficiencies in generation, transmission, and distribution through "no regrets" policies. Personal communication, Sanjeev Ahluwalia, February 2002.

69. Interview with international consultant, September 13, 2000.

REFERENCES

Ahluwalia, Sanjeev. 2000. "Power Sector Reforms: A Review of the Process and an Evaluation of the Outcomes." Refereed paper for the NCAER-ADB supported project on Economic Reforms. March.

———. 2001. "Power Experts Get it Wrong." *Economic Times.* June 5.

Ahluwalia, Sanjeev and Gaurav Bhatiani. 2000. "Tariff Setting in the Electric Power Sector: Base Paper on India Case Study." Paper presented at TERI Conference on Regulation in Infrastructure Services, November 14-15.

Balakrishna, P. 2000. "Power Regulators Grapple with Tariff Issues." July. Online at: http://indiapoweronline.com (March 29, 2002).

Bulletin on Energy Efficiency. 2002. "Preferential Tariff for Wind Energy Proposed." *Bulletin on Energy Efficiency.* 2 (4) February. Online at: www.renewingindia.org (March 29, 2002).

———. 2002. "Strategic Approach and Road Map for the Establishment of the Bureau of Energy Efficiency." *Bulletin on Energy Efficiency.* 2 (4) February. Online at: www.renewingindia.org (March 29, 2002).

Churchill, A.A. and R.J. Saunders. 1989. "Financing of the Energy Sector in Developing Countries." Fourteenth Congress of the World Energy Conference, Montreal, Canada, September 14-22.

Desai, Ashok. 1999. "The Economics and Politics of Transition to an Open Market Economy: India." OECD Technical Paper. CD/DOC (99) 12.

Dixit, S., G. Sant, and S. Wagle. 1998. "Regulation in the WB-Orissa Model : Cure Worse Than Disease." *Economic and Political Weekly.* XXXIII (17).

Dubash, Navroz K. 2002. *Tubewell Capitalism: Groundwater Development and Agrarian Change in Gujarat.* New Delhi: Oxford University Press.

Echeverri-Gent. 2000. "Politicians' Incentives to Reform: Globalization, Partisan Competition, and the Paradox of India's Economic Reform." Unpublished paper. Online at: http://www.people.virginia.edu/~jee8p/smuart2.htm (April 18, 2002).

Economic Times. 2001. "Government Rules out Spot Market for Power Trading." *The Economic Times.* (December 6).

Fernandes, D. and L. Saldanha. 2000. "Deep Politics, Liberalisation and Corruption: The Mangalore Power Company Controversy." Law, Social Justice and Global Development. 1. Online at: http://elj.warwick.ac.uk/global/issue/2000-1/fernandes.html (April 18, 2002).

Government of India, 1980. "Report of the Committee on Power, (Chaired by V.G. Rajyadaksha), Ministry of Energy, Department of Power." New Delhi: Government of India.

———. 1991. "Policy on Private Participation in Power Sector." Gazette of India, October 22. No. 237. New Delhi: Ministry of Power. Online at: http://www.indiainfoline.com/infr/lapo/poli/popv.html (April 18, 2002).

———. 1999. "Power Sector Reform and Restructuring–Financial Restructuring of GRIDCO, Orissa–Interim proposals." November 29. New Delhi: Ministry of Power.

———. 2001. "Conference of Chief Minsters/Power Ministers: Agenda Notes." March 3.

Gutiérrez, L. 1993. "International Power Sector Experience: A Comparison with the Indian Power Sector." Conference on Power Sector Reforms in India, Jaipur, India, October 29-31, Energy Sector Management Assistance Program, Washington, D.C.

Indiapoweronline.com. 2000a. "Power Regulators Grapple with Tariff Issues." July. Online at: http://indiapoweronline.com (April 4, 2001).

———. 2000b. "Regulators Get Down to Business." September. Online at: http://indiapoweronline.com (April 4, 2001).

———. 2001a. "Orissa: What Went Wrong?" July. Online at: http://indiapoweronline.com (April 4, 2001).

———. 2001b. "Expert Group Recommends Securitisation Contingent on Reform." June. Online at: http://indiapoweronline.com (April 4, 2001).

Mahalingam, S. 2000. "Power Reforms in Trouble." *Frontline.* 17 (March 4-17): 94-98.

———. 2002. "The Great Power Robbery." *Frontline.* 19 (March 16-29).

Menon, Parvathi. 2002. "Undistributed Risks." *Frontline.* 19 (February 16-March 1).

Pillai, S.M.C. and R. Krishnamurthy. 1997. "Problems and Prospects of Privatization and Regulation in India's Power Sector." *Energy for Sustainable Development.* III(6):67-76.

Planning Commission of India. 1994. "Report of the NDC Committee on Power." New Delhi: Government of India.

Pragativadi.com. 2002. "Finding working capital not mandatory: BSES." Online at: http://www.pragativadi.com/260102/business.htm (April 4, 2002).

Purkayastha, Prabir. 2001. "Power Sector Policies and New Electricity Bill: From Crisis to Disaster." *Economic and Political Weekly.* XXXVI (June 23): 2257-2262.

Rajan, A. Thillai. 2000. "Power Sector Reform in Orissa: An Ex-Post Analysis of the Causal Factors." *Energy Policy.* 28: 657-669.

Reddy, A.K.N., G.D. Sumithra, P. Balachandra, and A. D'Sa. 1991. "A Development-focused End-use-oriented Electricity Scenario For Karnataka." Parts I and II. *Economic and Political Weekly.* XXVI (14 & 15): 891-910 and 983-1001.

Reddy, A.K.N. and A. D'Sa. 1995. "The Enron and Other Similar Deals vs. the New Energy Paradigm." *Economic and Political Weekly.* XXX (June 17):1441-1448.

Reddy, A.K.N. and Sumithra, G. 1997. "Karnataka's Power Sector—Some Revelations." *Economic and Political Weekly.* XXXII (12): 585-600.

Sankar, T.L., and U. Ramachandra. 2000. "Electricity Tariffs Regulators: The Orissa Experience." *Economic and Political Weekly.* XXXV May 27 (21/22):1825-1834.

Sant, G. , and S. Dixit. 1996. "Beneficiaries of IPS Subsidy." *Economic and Political Weekly.* XXXI (December 21).

Suri, Jyoti. 2000. "Attempt at Real Reforms" *Powerline.* April 10-16.

Tata Energy Research Institute. 1993. "Performance of the Power Sector in India." Conference on Power Sector Reforms in India, Jaipur, India, October 29-31, Energy Sector Management Assistance Program, Washington, D.C.

Vergin, Heinz. 1993. Letter to M.S. Ahluwalia, Secretary, Ministry of Finance, Government of India, April 30.

Verma, P.S. 1999. "Akali-BJP Debacle in Punjab Wages of Non-Performance and Fragmentation." *Economic and Political Weekly.* XXXIV (December 11-17).

World Bank. 1991. *India: Long Term Issues in the Power Sector.* Washington, D.C.

———. 1993a. *The World Bank's Role in the Power Sector.* Washington, D.C.

———. 1993b. Conference on Power Sector Reforms in Jaipur, India, October 29-31, Energy Sector Management Assistance Program, Washington, D.C.

———. 1996. "Staff Appraisal Report: Orissa Power Sector Restructuring Project." Report No. 14298 (April 19).

———. 1997a. "Inspection Panel Report and Recommendation, NTPC Power Generation Project." Loan 3632-IN (July 24).

———. 1997b. "Haryana Power Sector Restructuring Project." Project Appraisal Document. Report No. 17234 (December 16).

———. 1998. "India: Environmental Issues in the Power Sector." Report No. 205/98 (June).

———. 1999a. "Meeting India's Energy Needs (1978 - 1999): A Country Sector Review." Report No. 19972. Vol. I (December 23).

———. 1999b. "Andhra Pradesh Power Sector Restructuring Project." Project Appraisal Document. (January 25).

5

INDONESIA

ELECTRICITY REFORM UNDER ECONOMIC CRISIS[1]

Frances Seymour[2]
Agus P. Sari[3]

INTRODUCTION

By mid-2001, when Megawati Sukarnoputri assumed the presidency of Indonesia, electricity sector reform had not progressed far. A brief spurt of reform-oriented activity in the aftermath of the 1997–98 Asian financial crisis lost momentum as the country failed to regain economic and political stability. Nevertheless, the Indonesian experience usefully illuminates constraints on the inclusion of environmental objectives and other elements of the public benefits agenda in electricity sector reform, as well as limitations on the leadership of international donor agencies in this arena.

BACKGROUND

The Electricity Sector in Indonesia[4]

As in many other countries, the contours of the electricity sector in Indonesia have been shaped in part by the country's history, geography, and natural resource endowment. Although there had been private commercial production of electricity during the Dutch colonial period and briefly following independence in 1945, the national government has taken the lead role in the development and administration of the electricity sector in Indonesia for the last half-century (Pelangi, 2000). The National Electric Power Company (*Perusahaan Umum Lisktrik Negara*, hereafter PLN) was established in 1950, and has been a key player in the rapid development of the sector (GOI, 1998a). By the 1990s, PLN was one of

the largest such companies in the world, with some 22 million customers and more than 50,000 employees (Pelangi, 2000). Box 5.1 provides a profile of the electricity sector in Indonesia.

BOX 5.1 | **PROFILE OF THE ELECTRICITY SECTOR IN INDONESIA**

Population (2000)[1] : 212 million

Households connected to PLN (1996)[2]:
Total: 67%   *Rural: 51%*   *Urban: 90%*

Installed electricity generation capacity (1999)[3]
Total: 21 gigawatts (0.7% of total world capacity)
> *Thermal: 81%*
> *Hydro: 14%*
> *Nuclear: 0*
> *Geothermal and Other: < 2%*

CO_2 emissions from electricity and heat as a share of national emissions (2000)[4]: 21%

Notes:

1. World Resources Institute. 2000. *People and Ecosystems: The Fraying Web of Life*. Washington, D.C.: World Resources Institute.

2. Asian Development Bank (1999b).

3. www.eia.doe.gov/pub/international/ieapdf/t06_04.pdf (February 6, 2002).

4. Computed by WRI using International Energy Agency (IEA) data. IEA, 2001. *CO2 Emissions from Fossil Fuel Combustion*. Paris: OECD.

PLN's installed capacity represents only 58 percent of Indonesia's total; most of the rest is captive power for the manufacturing industry. (Captive power is off-grid generation capacity dedicated to a specific electricity requirement.) Captive power has been installed mainly by facilities without easy access to PLN's distribution grid—almost half is outside Java and Bali—or to provide backup for PLN's unreliable service (Kristov, 1995). Approximately 60 percent of captive power capacity is estimated to come from diesel generators, and about a quarter from cogeneration plants. While the diesel-fueled portion has negative environmental impacts, captive power may be a more efficient alternative to PLN supply for companies using process steam, and for the high peak loads required by extractive industries in the outer islands.

In PLN's Java-Bali system, which has been the focus of restructuring efforts and is the subject of this chapter, gas and coal are the dominant sources of electricity. Gas combined-cycle generation and coal-fired steam constituted about 6 gigawatts and 6.5 gigawatts, respectively, or about 85 percent of the total in 1998. Gas and coal are expected to continue to be the main sources of electricity in the Java-Bali system, as large hydropower projects are believed to have reached peak energy production (PLN, 1998). While hydrocarbon fuels have been heavily subsidized in Indonesia, the government has also promoted increased use of the country's 20,000 megawatt geothermal potential as a renewable alternative to fossil fuels.[5]

Electricity development in Indonesia has experienced extraordinarily high growth rates over the last 20 years. PLN's installed capacity grew at a rate of 15 percent per year between 1982 and 1989 (Pape, 1999), and overall growth continued at 10 percent per year between 1990 and 1998 (PLN, 2000). Even during the financial crisis in 1998 (described below), when the economy experienced negative growth of 15 percent, growth in the electricity sector continued at 4 percent per year. In light of Indonesia's 55 percent electrification rate (80 percent on Java and Bali, 20 percent on other islands), there is still considerable room for continued growth. The Indonesian government projects demand to increase by 8.9 percent per year between 2000 and 2010 in Java and Bali, and 10 percent per year outside Java and Bali (GOI, 2000c).

In addition to being an important driver and a reflection of Indonesia's economic development, the electricity sector also plays significant social and political roles. Provision of electricity at low uniform rates to consumers throughout the archipelago has symbolized a commitment to social equity within and among the islands. As in other countries, populist sentiment has proven a political constraint on raising electricity tariffs. Maintenance of low uniform rates has required cross-subsidies within the electricity sector, as well as between the electricity sector and the national budget. These cross-subsidies have not been transparent.

The Special Relationship with the World Bank

The World Bank's involvement in the electricity sector in the 1980s was emblematic of the Bank's special relationship with the Government of Indonesia during that period.[6] The oil crisis in the 1970s had focused the government's attention on the need for structural reform in the sector, which was based on provincial oil-fired steam plants and diesel-based self-generation, in order to release those fuels to the lucrative export market.[7] In the 1980s, Indonesia was the Bank's largest borrower in the electricity sector; by mid-1989, the Bank had financed 18 projects in the sector (World Bank, 1996). As a complement to this large loan portfolio, the Bank also produced analytical reports (so-called "Economic and Sector Work," or ESW) on the Indonesian electricity sector at an unusually high rate, averaging one per year from 1981 to 1988 (World Bank, 1989).

World Bank staff worked closely with technocrats in the government to pursue a joint long-term agenda of investment and reform to integrate the system and expand capacity based on coal and hydropower. This collaboration was characterized by one Bank official as a "model relationship with a borrower" in the context of a "golden age at the dawn of Indonesian reform."[8] However, other analysts have noted that the Bank was not successful in promoting price reforms during this period. Indeed, with the exception of a brief suspension of lending in 1987–88, the Bank continued to support the sector despite "long periods of noncompliance even as to its financial covenants" (Haugland, et al., 1997).

The Bank's ability to exercise influence through the persuasiveness of its analysis and its partnership with the technocrats was demonstrated in 1987, when a power struggle came to a head inside the government over whether or not to invest in nuclear power. Future President Habibie, then Minister of Research and Technology, was the main proponent of the nuclear option. His case was bolstered by a parade of Western heads of state who visited Jakarta to promote agreements on technical cooperation designed to generate business for Western corporations. While these governments and corporations and their partners in the Indonesian nuclear agency produced massive studies in favor of nuclear power, opponents in the Ministry of Finance were able to prevail by utilizing a modest analysis by the World Bank showing the high cost of nuclear compared to coal.[9]

Many of the issues that would arise in the context of electricity sector restructuring in the late 1990s were foreshadowed in the Bank's lending and policy dialogue with Indonesia's government in the 1980s. Most importantly, the joint agenda of the Bank and the technocrats focused on the corporatization of PLN. A 1989 Power Sector Institutional Development Review recommended that, in order to meet the demand for rapid expansion, PLN should pursue a strategy of "deregulation, decentralization, and competition" in order to move "from bureaucracy to enterprise" (World Bank, 1989).

The impetus for these restructuring prescriptions was the need to attract private capital to finance the growth in generation capacity necessary to meet soaring demand for electricity.[10] Box 5.2 provides a chronology of electricity restructuring in Indonesia.

The 1989 sector review suggested the possibility of breaking up PLN into smaller units that might eventually be candidates for privatization. In addition, the report suggested that the government consider creating an environment in which private power producers could compete with PLN. The report was prescient, however, in cautioning that while privately owned power plants were a potentially attractive option, "their economic advantages for Indonesia cannot be taken for granted and need to be evaluated with care" (World Bank, 1989).

The Era of Independent Power Producers

Financing of the Paiton Thermal Power Project in 1989 marked the beginning of the end of the World Bank's special relationship with the Indonesian government in the power sector, in large part because it coincided with the advent of Indonesia's experiment with private participation in the sector. While the project itself was satisfactorily completed in 1995, two issues arose during its implementation that would strain cooperation to its limits in the late 1990s. First was the issue of the corruption that was endemic to private power deals negotiated in the absence of competition. This concern appears

CHRONOLOGY OF ELECTRICITY SECTOR REFORM IN INDONESIA

1985	New Electricity Law passed.
1989	World Bank sector review recommends introduction of competition and possible eventual privatization.
1990	President Suharto approves first Independent Power Producer (IPP) project.
1992	Implementing regulations for 1985 law promulgated as Presidential Decree No. 37, which encouraged private participation in the sector.
1994	Government Regulation No. 23 corporatizes PLN.
1994–1997	25 additional IPP projects signed.
1997	Asian financial crisis sweeps Indonesia, bankrupting PLN.
January 1998	World Bank suspends new lending to the electricity sector.
May 1998	Civil unrest—in part driven by tariff increases—forces President Suharto to step down.
August 1998	Habibie government announces electricity sector restructuring policy, issuing a "White Paper" following a workshop with donors.
March 1999	Asian Development Bank (ADB) announces $400 million in loans to support Indonesia's electricity sector restructuring program; Japan Bank for International Cooperation follows with a $400 million loan tied to the ADB support.
October 1999	Indonesia's first democratic elections replace Habibie with Abdurrahman Wahid.
February 2000	Controversy erupts in Parliament and in the press over proposed tariff increases.
February 2001	Government forwards new draft electricity legislation to Parliament.
August 2001	Adburrahman Wahid is replaced by Megawati Sukarnoputri.
October 2001	Parliament passes new oil and gas law.
November 2001	Public hearing on draft electricity law held by House of Representatives.

between the lines of the project's implementation completion report prepared by the World Bank:

It is interesting to note that, for contracts awarded through competitive bidding, the actual costs were slightly less than the estimated costs for all contracts, whereas for contracts awarded through direct negotiation, such as in the case of bilateral cofinancing, the actual costs came out to be significantly higher than the estimated costs (World Bank, 1996).

Second was the issue of excess generating capacity on Java, a prospective problem that would be exacerbated by a whole generation of independent power projects initiated in the early 1990s.

The door had been opened to private sector participation in electricity generation by a law passed in 1985. However, the law only came into effect when the necessary accompanying regulations were promulgated through Presidential Decree No. 37 in 1992, which encouraged the participation of private enterprises in electricity generation, transmission, and distribution. An unfortunate feature of the decree was that it opened the door to unsolicited proposals for the private production of electricity.[11] At the time, the privatization of electricity generation was opposed by the head of PLN's research division, who argued that privately supplied electricity would be almost 50 percent higher than PLN's costs due to equity return requirements and interest rates (Sudja, 1993). However, a later study found that if all hidden subsidies were taken into account, PLN's true generation costs would increase by 46 percent (Kristov, 1995).

Prior to the 1992 decree's enactment, then-President Suharto in 1990 had agreed to develop another coal-fired power plant at the Paiton site as the first private power project in Indonesia. This was soon followed by another. Because the new plants reached financial closure before the World Bank-supported plant at the site began operations, and were financed under "take-or-pay" contracts, PLN would be forced to utilize their capacity in preference to power from the first plant, thereby undermining the economics of the Bank's investment (World Bank, 1996). The World Bank and an advisor from USAID had counseled the Government of Indonesia to "start small" in its experimentation with private power production to reduce risk. Future President Habibie, however, wanted to "start big," and—in contrast to the fate of his nuclear ambitions in the 1980s—got his way with the two plants at the Paiton site. The commercial interests of donor governments aided his case. U.S. President Clinton's visit to Indonesia in late 1994 for the Asia-Pacific Economic Cooperation (APEC) meeting put pressure on both governments to have deals ready for his signature. Clearly, the influence of the World Bank and its technocratic allies in the government had been eclipsed by other actors.

During 1994–97, 25 more power-purchase agreements (PPAs) were issued and signed with independent power producers (IPPs). The majority of these agreements were based on unsolicited, nontransparent bidding processes, and resulted in overpriced, dollar-pegged, take-or-pay conditions that greatly favored project investors. These projects were driven by the interests of private developers, who often had close connections to the president's family and cronies. Many were also linked with North American, European, and Japanese corporations, which were in turn backed by bilateral export credit and guarantee agencies.[12] The level of corruption seen in these deals has been characterized as "staggering" (Fried, 2000).

The World Bank had counseled the Government of Indonesia to "start small" in its experimentation with private power production to reduce risk.

The PPAs for new installed capacity did not reflect PLN's long-term plans for development of generation and transmission capacity. Indeed, many of the new plants would produce unneeded power, sometimes far from existing transmission lines. The World Bank had first expressed concerns about the "looming problem" of excess capacity in late 1993, even before the surge of unneeded PPAs. Bank officials subsequently "raised this issue repeatedly at high levels of authority in its continuing policy dialogue with GOI" (World Bank, 1996), including forceful representations by the Bank's country director in mid-1994.[13] In November of that year, the Bank sent a strongly worded letter warning of the prospective $8 billion dollar cost—"a staggering figure"—that PLN would incur over the next 10 years from excess capacity and excessive costs. Bringing down the projected excess capacity reserve margin—one of the highest in the world—was made a condition of the Second Power Transmission and Distribution Project in 1996 (World Bank, 1996). In 1995, the World Bank's Country Economic Memorandum for Indonesia warned that the country risked following the example of the Philippines, in which the national utility was forced to purchase power from high-cost take-or-pay contracts with private producers while reducing use of its own lower-cost generation capacity (World Bank, 1995). Box 5.3 describes the Philippine experience.

Financial Crisis and Political Change

In mid-1997, the Asian financial crisis swept through Indonesia. The Indonesian currency, the Rupiah, lost 80 percent of its value in only four months. Prices soared, capital fled the country, factories closed, and within a year some 30 million people fell below the poverty line, joining the 26 million people already there (World Bank, 2001b). Unlike many developing countries Indonesia had, until 1997, escaped the

institutions ($300 million from the Asian Development Bank, and $400 million each from the World Bank and Japan's Bank for International Cooperation). But the release of tranches of these financial packages was conditioned on the passage of legislation that would privatize Napocor, vertically unbundle the electricity sector into competitive or regulated electricity markets, and eliminate cross-subsidies among consumer groups—as well as improvements in Napocor's financial performance.

The question of what institutions or economic groups would assume responsibility for Napocor's stranded debts under the proposed reforms created widespread political controversy in 1999 and 2000. Domestic constituencies critiqued the GOP's proposed solution contained in legislation under consideration before the Philippine Congress: imposition of a universal levy on electricity consumers. The PPAs were to be transferred to buyers of Napocor's generating assets, but the government would cover the cost differential between the price to be paid under the PPAs and the market price of power. This burden would also be transferred to electricity consumers in the proposed levy.

These controversies and the political crisis generated by the impeachment of President Estrada in 2000 stalled the passage of the reform legislation championed by international donors. The election of Gloria Macapagal-Arroyo in May 2001, and a renewed effort by Arroyo's government to build a domestic constituency for the passage of reform legislation broke the deadlock. Domestic constituency-building was assisted by a stakeholder

consultation process supported by the USAID mission in Manila. In June 2001, the Philippine Senate and House of Representatives passed the Electric Industry Reform Act (EIRA). The act preserved the original reform program supported by donors, but included some key concessions to domestic constituencies, such as a 5-percent cut in electricity tariffs for the poorest populations and recognition that PPAs would need to be renegotiated.

The Philippine experience demonstrates that private sector entry into the business of electricity generation prior to political consensus on structural and tariff reforms can impede the forging of such consensus.

Sources:

Asian Development Bank. 1998. "Report and Recommendation of the President to the Board of Directors on a Proposed Loan and Technical Assistance Grants to the Republic of the Philippines for the Power Restructuring Program." Report No. RRP: PHI 31216. (November).

Carpio, Denis T. 1998. "Power Industry Restructuring in the Philippines: Issues and Alternative Solutions." Conference paper. 17th Congress of the World Energy Council, Houston, Texas, September.

Dalangin, Lira. 2001. "Filipinos to Pay P619 Yearly for Napocor Debt: Work Group." Online at: http://www.inq7.net/brk/2001/may/26/brkpol_8-1.htm (May 31, 2002).

USAID. No date. Philippines Activity Data Sheet.

Winrock International. 2001. "Renewable Energy State of Industry." Reports No. 1 and 2. (May and August).

Wong, Susanne. No date. "An Overview of ADB's Support for Energy Sector Reform." Briefing Paper 10. International Rivers Network.

discipline of externally imposed stabilization and structural adjustment programs. However, in September, it was forced to request a bailout package organized by the International Monetary Fund (IMF).

The financial crisis caused chaos among the government agencies, private developers, financiers, and international financial institutions involved in the Indonesian power sector. PLN was plunging into bankruptcy. The value of revenues in Rupiah declined even as dollar-denominated debts, take-or-pay contracts, and spare-parts prices soared.

Projects that had been of questionable viability even before the crisis became even less defensible in light of a crashing economy and currency. Between July and December, a flurry of communications regarding projects in various stages of development ricocheted among the president's office, the Ministry of Mines and Energy, PLN, the World Bank, and the IMF. Presidential Decree No. 37/1997 subsequently suspended some 27 IPP projects.

Exacerbating concern that PLN would not be able to meet its existing financial obligations, much less new ones, was the proposed Tanjung Jati C power plant— an IPP project linked to President Suharto's daughter and steeped in allegations of corruption. In a November 1997 letter to the Ministry of Mines and Energy, the World Bank's country director suggested that the project be reconsidered. The signing of the deal was announced in December; in January the country director informed the ministry that it was suspending a loan then in preparation, thus terminating almost three decades of lending to the Indonesian electricity sector. The "special relationship" between the World Bank and the Indonesian electricity sector had come to an end.

Soon the economic crisis led to a political crisis as well. By March and April 1998, riots and protests were spreading. Some of the protests were sparked by price increases mandated by the IMF, including a hike in electricity tariffs in March. Following the shooting deaths of several student demonstrators, tens of thousands of students took to the streets, eventually taking over the Parliament building. Ultimately, President Suharto stepped down in May, and was replaced by his vice president, Habibie.

"REFORMASI" IN THE ELECTRICITY SECTOR

Historically, major changes in Indonesia's trade and industrial policy have been linked to major political and economic crises (Pangestu, 1996). In mid-1998,

it appeared that the political and economic events of the preceding months had set the stage for restructuring of the electricity sector. *Reformasi total* or "total reform" was the slogan on everyone's lips, and it appeared that radical changes in the way the country was governed were in progress. Electricity sector reform appeared inevitable, as PLN's financial viability had been destroyed by the combination of the economic crisis and the accumulation of ill-considered PPAs. Suddenly, the reform gained momentum, largely due to the need to privatize as many PLN components as possible to generate cash and staunch financial hemorrhage.

IMF Conditionality and Kuntoro's Leadership

In responding to the economic crisis, the electricity sector in particular was a focus for reform for both the government and the international donor community. In a series of letters of intent and supporting documents, the government made commitments to the IMF that included establishing the legal and regulatory framework to create a competitive electricity market; restructuring of the organization of PLN; adjustment of electricity tariffs; and rationalizing power purchases from PPAs (GOI, 1999a).

Habibie-promoted nuclear and IPP schemes had been antithetical to the World Bank's agenda in the electricity sector. Ironically, Habibie as president created conditions favorable to reform. Kuntoro Mangkusubroto, the Minister of Mines and Energy, was one among "as good a crop of ministers as you could hope for," according to one World Bank official.[14] Kuntoro quickly set about fulfilling Indonesia's commitments to the IMF to reform the electricity and oil and gas sectors. He took ownership of the results of an independent audit of PLN commissioned by the IMF that highlighted the need for improved efficiency, and pushed for the formulation and passage of new legislation to govern the sector.

Kuntoro himself took a personal interest in the design of electricity sector reforms and building constituencies for them. He is said to have carefully corrected and commented on five drafts of a "White Paper" that laid out the content of the proposed reforms.[15] While the white paper was under development, Kuntoro had periodic breakfast meetings with key stakeholders in the electricity sector. These included government officials, business people, and nongovernmental activists. Despite the exclusive appearance of "by-invitation" breakfasts with the minister, these meetings were among the first attempts to open up participation in electricity sector decisionmaking. The so-called "breakfast club" evolved into a formal organization—the Indonesian Electricity Society—to serve as a forum for exchange between the minister and stakeholder representatives.

In August 1998, the minister convened a major workshop attended by representatives of various government and donor agencies at which the draft white paper was discussed. The paper was then released to the public, and was later cited in a letter of intent to the IMF as the basis for the government's electricity sector restructuring policy (GOI, 1999a).

Content of the Proposed Reform[16]

The August 1998 white paper articulated four objectives for electricity sector restructuring: (1) the restoration of financial viability; (2) competition; (3) transparency; and (4) more efficient private sector participation. The six areas targeted for reform were (1) industry restructuring and unbundling; (2) introduction of competition; (3) tariff-setting, cost recovery, and removal of subsidies; (4) rationalization and expansion of private sector participation; (5) redefinition of the government's role; and (6) strengthening of the legal and regulatory framework (GOI, 1998a).

The white paper reflected a reform agenda informed by a series of studies commissioned by the World Bank or by the GOI in previous years. A study commissioned by the Bank from the consulting firm Norplan A/S in 1993 provided an institutional framework for restructuring, which included unbundling of the generation, transmission, and distribu-

tion functions of PLN (Norplan A/S, 1993). The GOI subsequently commissioned a study by Coopers and Lybrand on the regulatory framework and strategy for private power production that was completed in 1996.[17]

The restructuring agenda put forth in the white paper aims to separate the commercial, social, and regulatory functions of PLN. Government Regulation No. 23 of 1994 had already changed PLN's status from a public utility to a public company, marking the corporatization of PLN. In the restructured sector, electricity producers would operate commercially and be financially independent of the government. Social functions would be continued by the government, which would provide transparent subsidies to poor regions and customers from a Social Electric Power Development Fund. Regulatory roles would be played by a new, autonomous agency distinct from the Ministry of Mines and Energy to be established under a new electricity law to be enacted by 1999.

According to the 1998 plan, the independent regulatory body, an independent transmission company, and an independent regional company would begin operation in 2000, while subsidies were gradually removed. Generation and distribution companies would operate under the supervision of the Java-Bali Electricity Company (JBECs). A PLN services company would be the only remaining component of the former monopoly. It was expected that in 2001–2002, independent generation and distribution companies would begin to emerge, while the ones controlled by JBEC would either be privatized or enter into direct competition with the IPPs; and by 2003, a complete multiple buyer/multiple seller system would emerge, as all generation and distribution companies would be independent and JBEC would be privatized. Figure 5.1 compares the system in 1998 to the one envisioned for 2003.

Many details of the restructuring process were to be treated separately in three additional documents (GOI, 1998a). A tariff code, with the status of a presidential decree, would establish tariff structures and subsidy mechanisms. A ministerial-level plan-

portfolio of project lending in various sectors, the Bank now focused on a policy reform agenda leveraged through a series of policy reform support loans coordinated with the conditionalities of the IMF bailout package. Among the conditionalities prepared by World Bank staff were those related to the electricity sector. The World Bank was a key participant in a flurry of activities focused on electricity sector reform in mid-1998. In July, Bank officials participated in the drafting of the white paper mapping out the restructuring plan championed by Minister Kuntoro, and according to an international consultant involved in the process, were primarily responsible for the "public benefits" content that it contained.[19]

The Asian Development Bank (ADB) also played an increasingly important role. Like the World Bank, the ADB had been lending to the Indonesian electricity sector for almost three decades, with some 28 loans totaling more than $3 billion (ADB, 2001). The financial crisis marked a shift in the ADB's role from project financier to a proponent of policy reform backed by sector lending. At the time of the August 1998 workshop, an ADB mission was in Jakarta to begin preparation of an electricity sector loan. Indeed, according to an ADB official, the ADB had conditioned its assistance on production of the white paper.[20] In March 1999, the ADB's board approved a $380 million loan to support the government's restructuring agenda, and an additional $20 million technical assistance loan for "capacity-building to establish a competitive electricity market" (ADB, 1999a). Co-financing in the amount of $400 million was arranged from the Japanese Bank for International Cooperation.

The World Bank's suspension of new lending in the sector in early 1998 followed by the ADB's entry with a major sector loan prompted a shift in the relative influence of the two institutions over the reform agenda. The ADB and the World Bank worked out an informal division of labor in the electricity sector: ADB took the lead on sectoral restructuring issues, while the World Bank focused on PLN corporate and financial restructuring. Thus, the ADB would support technical assistance for development of the

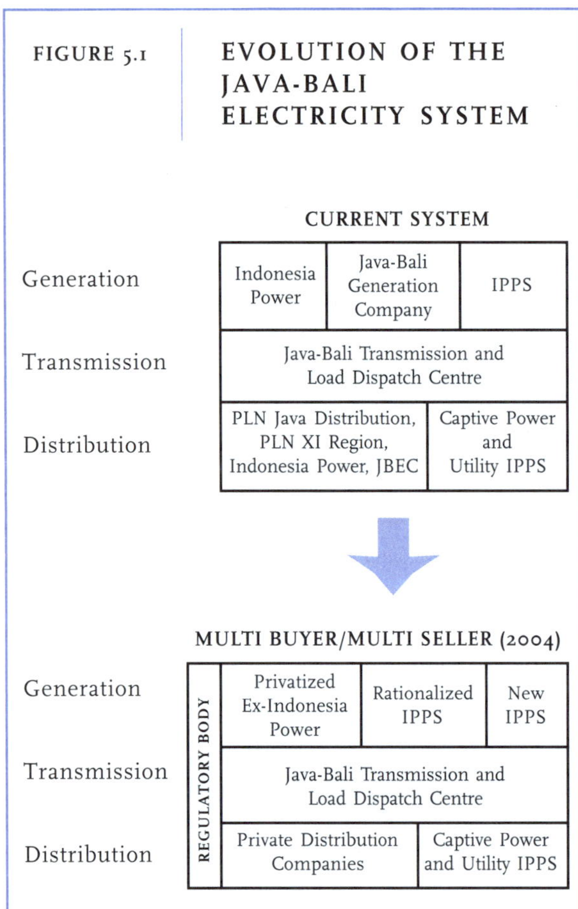

ning and competitive tendering code would detail planning arrangements and procedures for bidding on generation and transmission projects. Finally, a ministerial-level grid code would govern operation of power transmission, scheduling, and dispatch. The fate of much of the public benefits agenda would reside in these documents, which were being written by donor-funded consultants.[18]

The Role of International Donor Agencies

International donor agencies played important roles in the electricity sector reform process. The 1997–98 financial crisis transformed the World Bank's overall relationship with the government. Instead of a large

draft law, the tariff, procurement, and grid codes, as well as development of a new regulatory body.

One of the objectives of ADB support was to build public acceptance of the electricity price increases that would be necessary to restore PLN's financial viability, while protecting the interests of consumers (ADB, 1999a). Toward this agenda, the ADB loan supported assistance to the government and PLN in building the case for tariff increases, and the convening of a multistakeholder, Working Group on Restructuring, which brought together nongovernmental organizations, industry lobbyists, and government officials.

The United States Agency for International Development (USAID) played an interesting role in Indonesia's electricity sector restructuring process. Over the last decade, USAID has financed a series of experts to advise the Indonesian government on electricity sector issues. In 1990, a USAID-funded advisor based in Jakarta had joined the World Bank in cautioning the Indonesian government against its headlong rush into private participation in electricity production.[21] The expert's advice was reportedly an irritant to the U.S. ambassador, as it conflicted with the embassy's agenda of promoting U.S. commercial interests in the sector (ADB, 1999a).

A conflict of interest within the U.S. government regarding IPPs arose again after the financial crisis, when the Government of Indonesia canceled a PPA for a geothermal energy plant with California Energy. California Energy's investment had been guaranteed by the U.S. Overseas Private Investment Corporation (OPIC). After the company went to arbitration with a claim of $290 million and won, OPIC had to pay out $217.5 million and private insurers $72.5 million (Watson, 2001). As soon as the case entered arbitration, USAID had to withdraw an advisor providing legal advice to the government on how to deal with the IPP issue.[22] According to USAID officials in Jakarta, although PLN officials requested further assistance to help them deal with the issue, USAID had to say no. After protracted negotiations, the Indonesian government ultimately agreed to pay $260 million to reimburse OPIC and other insurers (Harvest International, 2000).

USAID was otherwise active in electricity sector restructuring debates, and the staff participated in the August 1998 workshop that finalized the white paper. The agency has also provided financial support for Indonesian NGOs such as the Indonesian Consumers Union, encouraging them to speak out in the restructuring process.[23]

Both USAID and the World Bank claim credit for promoting renewable energy in the context of power sector reform. According to the staff in Jakarta, USAID has championed a grid code that would include renewable energy sources without penalty.[24] Just before the financial crisis hit in mid-1997, the World Bank had approved a loan and companion grant from the Global Environment Facility to create a market for privately financed renewable energy projects (World Bank, 1997). Under the project, which was canceled in the wake of the financial crisis, PLN would have offered to purchase power from small-scale renewable sources at tariffs based on PLN's avoided costs.

The Process Stalls

With the electricity sector restructuring white paper and donor assistance in place, reform was expected to progress quickly. While continued reserve generation margins and the economic downturn provided some breathing space, there was concern that increasing demand for power would soon exceed available generating capacity, lending a sense of urgency to the process. Given the long lead time for new projects to come on line, reform advocates saw the post-crisis period as a window of opportunity to restructure the sector and attract new investment before power shortages began to constrain economic growth.[25] In a March 1999 letter of intent, the government articulated its expectation that a new electricity law would be passed by December 1999 (GOI, 1999a). However, a series of factors conspired to stall the process.

First, there is some indication that key officials in the Indonesian government chose to make reform of the oil and gas sector a higher priority. While the oil and gas sector and the electric power sector were inextricably linked through distorted pricing of fuels, the former was seen as more difficult to reform due to the higher financial stakes.[26] In addition, PLN was seen as more prepared for restructuring than the national oil company (Pertamina), so priority attention was given to the latter.[27]

In October 1999, the first democratically elected People's Consultative Assembly was convened and selected Abudurrahman Wahid as president. When a new cabinet was formed, Kuntoro lost his position as Minister of Mines and Energy. The new government expressed its intention to proceed with electricity sector reform, and in January 2000 agreed to speed up the restructuring of a number of state-owned corporations, including PLN (GOI, 2000a). But after Kuntoro's departure, little progress was made. According to an international consultant, the new leadership in the ministry was not supportive of the restructuring agenda,[28] while many PLN employees had resented Kuntoro's leadership on the issue.[29]

The most significant stumbling block was the difficulty the government faced in raising electricity tariffs, which, in May 1999, were still below PLN's production costs (GOI, 1999b). While arguably essential to restoring PLN's financial viability, attracting new investment, and providing resources to expand access to electricity, tariff increases were opposed by a number of constituencies. Within the ministry, key officials responsible for drafting the new electricity law harbored doubts about proposed tariff increases. According to an international consultant, the May 1997 riots "didn't do much to encourage people in government to make hard decisions."[30] As described below, proposed tariff increases would later prove controversial when the draft law was submitted to the Parliament. Tariff increases were opposed by students and the left-wing Peoples Democratic Party (*Suara Pembaruan*, 2000b).

Another obstacle to the reform process was the difficulty faced by PLN in negotiating payment terms with the IPPs (GOI, 2000b). As long as this looming financial liability remained unresolved, the restructuring process could not move forward. Yet another obstacle was the political distraction of controversy over President Wahid's leadership, which eventually led to his replacement by Megawati Sukarnoputri in August 2001.

Tariff Increases and the Public Interest

The relationship between tariff increases and the public interest has proven to be the most contested aspect of the electricity sector reform process in Indonesia. On one side, international donor agencies, some government officials, independent analysts, and some NGOs argue that tariff increases are essential to maintain the viability of the electricity sector.[31] They point to the economic distortions induced by low tariffs—including inefficiency—and the fact that subsidized rates have drained the public treasury while benefiting electricity customers who can pay the economic cost of electricity. They assert that the impact of tariff increases on the poor can be mitigated through a rate structure that maintains a high subsidy content for the smallest household consumers. Furthermore, higher tariffs are necessary to finance expansion of access to electricity, especially to rural areas.

The relationship between tariff increases and the public interest is the most contested aspect of the electricity sector reform in Indonesia.

On the other side, students and more populist-oriented NGOs have demonstrated against tariff increases on the grounds that they will impose further hardships on the poor.[32] They argued that even if poor households are shielded from the direct impact of electricity tariff increases, those increases will provoke a rise in the prices of other goods consumed by the poor, including *Sembako*, an abbreviation for "the nine staple goods." Opposition to electricity tariff increases is part of a broader opposition to liberalization of the Indonesian economy that is driven by pressure from the IMF, the World Bank, and the Asian Development Bank.[33]

In February 2000, a national controversy erupted when a group of NGOs was reported to have supported a proposed 55-percent increase as part of a package discussed in the ADB-supported Working Group. The NGOs were excoriated in the press, described as being "biased against the people" (*Suara Karya*, 2000), and even accused of taking bribes.[34] In March, Kuntoro, now president of PLN, was the target of student protests against electricity price increases (Rakyat Merdeka, 2000).

Electricity tariffs were in fact increased in 2001, as subsidies to business and industrial users were gradually withdrawn, and further staged increases were planned for 2002. The debate continued into early 2002, as the new electricity law made its way through the Parliament. While the 2001 increases did not spark significant protest, small rallies against the higher 2002 tariff increases were staged in January in several cities by such groups as the Peoples Democratic Party (PRD) and the Anti

Subsidy-Revocation Alliance (A2PS) (*Tempo Interactif*, 2002b). Other groups, such as the Indonesian Consumers Foundation, were cautiously supportive of tariff increases, as long as they were accompanied by improvements in PLN's performance, and well-supervised implementation of the subsidy scheme for poor households (*Tempo Interactif*, 2002a).

THE PUBLIC BENEFITS AGENDA

The public benefits agenda—specifically concerns regarding social equity, environmental protection, and good governance—received uneven attention in the design of Indonesia's electricity sector restructuring process.

Concern about Social Equity

Indonesia's electricity sector restructuring agenda has focused almost exclusively on Java and Bali. The August 1998 white paper justifies this focus, suggesting that the Java-Bali electricity system is sufficiently mature for commercialization, while the smaller, more isolated systems on other islands, characterized by higher costs and lower electrification rates, should be restructured more slowly (GOI, 1998a). Mobilization of private resources for the sector on Java and Bali would enable the government to focus limited public finance on other islands— "where it is most needed" (GOI, 1998a).

At the time the white paper was being drafted, there was strong consensus among all parties about the need to protect social equity as part of the restructuring process. All participants were aware of the symbolic importance of affordable electricity in Indonesia historically. For example, in its sector review 10 years earlier, the World Bank had recognized the government's social objectives related to "equality"—including the provision of electrification throughout the archipelago—and "fairness," including the maintenance of electricity rates at affordable levels (World Bank, 1989). The white paper—and subsequently the draft electricity law—proposed a "Social Electric Power Development Fund" to finance continued subsidies to poor households and expanded access to electricity in underdeveloped regions.

In 1998, however, participants were also keenly aware that protests against price increases had helped bring down the Suharto regime. Concerns over the impacts of restructuring on the poor motivated the most significant sources of opposition to reform. Populist sentiments, expressed by student demonstrators and the popular press, denounced proposed tariff increases, and muted the voices of those NGOs who understood that the preponderance of existing subsidies were captured by the nonpoor at the expense of the national development budget.

Another source of opposition to reform related to its likely impact on employment, including from PLN's own labor union. The union may have feared job losses in the process of unbundling and after PLN was exposed to competition, but did not raise this concern in their public statements. Instead, the union criticized the proposed restructuring as unconstitutional. Article 33 of the Indonesian Constitution states that "branches of production that have large implications to the life of the general public should come under the control of the state." Unions have interpreted this provision to mean government ownership. Representatives of labor interests have been assertive participants in a civil society coalition for the Indonesian power sector formed in late 2001.

Industry associations also raised the specter of job loss as a reason to oppose electricity price increases. For example, in early 2000, the Indonesian Textile Association, though not an energy-intensive industry, threatened that increased tariffs would cause layoffs. Their lobbying efforts directed at the Ministry of Trade and Industry, the Ministry of Labor, and President Wahid were soon followed by similar special pleading from the Chamber of Commerce, the Indonesian Railway Company, and the hotel industry (*Bisnis Indonesia*, 2000a, 2000b; *Suara Pembaruan*, 2000a; *Republika*, 2000).

Proponents of reform, particularly in the international donor community, were frustrated by the public discourse on the impacts of reform on the poor and on employment, which according to their analyses was misleading. For example, the Poverty Impact Assessment conducted in conjunction with the ADB loan asserted that "the overall impact of the program is pro-poor" (ADB, 1999b). While conceding that tariff increases and restructuring would have short-term negative impacts (increases in prices of goods and services purchased by the poor and contraction in employment), the assessment points to targeted social protection mechanisms to deal with short-term impacts, and to lower prices and new employment opportunities in the long run (ADB, 1999b). A study undertaken by the Bandung Technology Institute indicates that the poorest households are willing to accept a 10-percent increase in electricity tariffs, while other electricity consumers are able to pay the full economic cost of power (Center for Research on Energy, 2001).

Populists denounced proposed tariff increases and muted the voices of those NGOs who understood that subsidies were captured by the nonpoor.

On the other hand, international NGOs have pointed out that the positive impacts promised by the donor agencies depend crucially on a set of assumptions about how the reforms will be carried out (Motoyama and Widagdo, 1999). In particular, they have questioned the availability of financing for the Social Electric Power Development Fund, and the accountability of the proposed independent regulatory board. (Motoyama and Widago, 1999)

The ADB's Poverty Impact Assessment also points out that only 40 percent of Indonesia's poor are connected to the grid. Thus, those who opposed tariff increases in the name of the poor failed to address the question of why continued subsidies to households already connected to the grid was a higher priority for public finance than expanded access to electricity for the poor majority.

Concern about Environmental Sustainability

In contrast to the first-order concern about social equity, concern about environmental sustainability was at best a third-order consideration in the design of Indonesia's electricity sector reform. Historically, the World Bank had worked with PLN to promote environmental protection at the project level. The 1989 sector review noted "environmental preservation" as among the objectives PLN had been asked to pursue. According to World Bank staff, PLN's performance in addressing the environmental and social impacts of its projects in the 1980s was "a shining model of how it should be done," pioneering best practices in resettlement for hydropower projects, and mitigation of emissions from coal plants.[35] The Paiton Thermal Power Project—financed the same year the World Bank formalized an environmental assessment policy in 1989—included among its objectives strengthening "the government's environmental monitoring capability as well as its ability to formulate and enforce environmental standards in the energy sector." According to the project's implementation completion report, significant results were achieved (World Bank, 1996).

However, environmental concerns do not appear to have influenced the design of electricity sector restructuring. The lack of attention to incentives for energy efficiency (other than tariff increases) is particularly striking in light of the potential for efficiency to reduce requirements for new generation capacity.[36] The World Bank's analysis focused instead on the environmental benefits to be realized from deregulating energy prices overall, which it was pursuing through proposed reforms in the oil and gas sector. Such deregulation would reduce subsidies to petroleum products used in electricity generation, such as diesel and fuel oil, and provide incentives for efficiency and for switching to cleaner fuels such as natural gas (World Bank, 2000).[37]

The August 1998 paper makes brief references to incentives for energy efficiency in tariff-setting as the last of six objectives (led by cost-recovery) under that section of the paper. The white paper makes no reference to incentives for the development of renewable resources, and no mention of environmental objectives per se. Similarly, while the preamble to the draft electricity law alluded to the need "to take into account the preservation of environmental functions, energy conservation, and energy diversification," only one sentence in the bill referred to the environment (Supplemental State Gazette, 2000).[38]

As described above, some donors and NGOs promoted increased attention to renewable resources in discussions of electricity sector reform. For example, in late 2001, an Indonesian environmental NGO hosted a regional meeting to discuss equity and sustainability in the context of electricity sector reform, at which participants concluded that insufficient attention was being given to environmental impacts and to removing barriers to renewable energy technologies (SPENA Newsletter, 2001). Others have questioned the lack of attention to demand side management in the reform process (Motoyama and Widagdo, 1999).

Concern about Good Governance

In the earliest discussions of power sector reform in Indonesia, attention to good governance included calls for increased financial transparency in the various subsidized transactions between PLN, the government, and consumers (World Bank, 1989). However, the focus of the reform effort in the late 1980s and early 1990s was on narrowing, rather than broadening, participation in decisionmaking. As part of the corporatization process, the World Bank's 1989 sector review recommended streamlining decisionmaking to reduce the "micromanagement" of PLN by government officials (World Bank, 1989). At that time, neither other ministries—such as the Ministry of Environment—nor civil society actors were on the radar screen as potential contributors to the process of reform.

By the late 1990s, attention to good governance in the Indonesia power sector restructuring process continued to focus on increasing the financial transparency of the sector, and added concern about stakeholder participation in the reform process itself. Little attention was devoted to the challenge of creating a new independent regulatory body to govern private participation in the electricity market.

In the era of *reformasi*, one thing that government, donor, and civil society actors could agree on was the need for increased transparency—one of the four objectives of power sector restructuring articulated in the white paper. Among World Bank/IMF conditionalities was a requirement that the government undertake an independent audit of PLN (GOI, 1998b), and NGOs and the Parliament demanded that the results of the audit be made public (*Media Indonesia*, 2000). Concerns raised by the IPP deals had put anti-corruption efforts high on the NGO advocacy agenda.

Good governance in the sector restructuring process focused on the financial transparency of the sector, and stakeholder participation in the reform process.

There was also an attempt to increase stakeholder participation in the reform process itself. As mentioned above, Minister Kuntoro conducted his own outreach early in the process through the "breakfast club." Later, a $20 million component of the ADB loan was allocated to develop constituencies for power sector reform through public awareness-building and engagement in the process. Accordingly, a multistakeholder, Working Group on Power Sector Restructuring, was set up. International NGOs questioned the ADB's emphasis on gaining acceptance of policies rather than civil society participation in the policymaking process (Motoyama and Widagdo, 1999). In addition, participation in an ADB-sponsored forum was controversial within the Indonesian NGO community. Due to a general perception that the ADB was not an institution working in the interest of Indonesia's long-term sustainable development, and a particular unwilling-

ness to be associated with the expenditure of loan funds that would add to Indonesia's debt burden, several key NGOs refused to participate.[39]

Perhaps due to the slow pace of reform, as of early 2002, participants in the reform process had paid little attention to the governance challenge of developing an independent regulatory function. The white paper envisioned that an autonomous agency reporting to the Minister of Mines and Energy would be created by the new electricity law. The new agency would be vested with regulatory authority for the entire energy sector, including electricity, gas, and oil. It would issue licenses and supervise compliance. While donor agencies saw such an autonomous agency as likely to be more transparent and accountable than its precursors, public interest advocates remained unconvinced in the absence of specifics (Motoyama and Widagdo, 1999).

The draft electricity law stated that "the Regulating Board shall make decisions independently and make a transparent clarification regarding all of the considerations in its decisionmaking," but deferred specifics to regulations that would be developed later. Prior to passage of the law and under pressure from investors, the government enacted Presidential Decree No. 15/2002, which revokes the 1997 decree that suspended IPP projects in preparation. How the government performs in supervising these projects will be an early test of commitment to genuine reform.

Given the recent history of corruption and political interference in the sector, in addition to the technical and managerial capacity that will be required to regulate the complexity of the proposed system, this governance challenge appears to have been underappreciated in the process to date.

CONCLUSION

The Indonesian experience with power sector restructuring was profoundly influenced by several contextual factors. The 1997 financial crisis and ongoing economic crisis both stimulated and complicated power sector restructuring. On the one hand, the collapse of the currency bankrupted the national electric utility; on the other hand, it made constituencies for labor (PLN employees union) and consumers (students and NGOs) more likely to oppose any reform that implied job losses or price increases.

At the same time, the legacy of independent power producers and purchase agreements strongly colored the domestic and international politics of power sector restructuring in Indonesia. Attention to the power sector on the part of domestic and international NGOs focused on the high degree of corruption associated with the agreements, and the likelihood that Indonesian consumers and taxpayers would end up shouldering the public and private debt incurred. The deep involvement of bilateral export credit and guarantee agencies in tainted deals constrained the ability of institutions such as the World Bank and USAID from providing technical assistance to the government to help remove this barrier to restructuring.

The 1998-99 political transitions created a context in which political roles and constituencies were in flux. The climate of *reformasi total* that followed the fall of Suharto opened political space for attention to increased transparency and NGO participation in government decisionmaking, as well as an increasingly meaningful role for the Parliament. However, a charismatic Minister of Mines and Energy who championed restructuring in the Habibie cabinet was replaced following Wahid's election.

The international donor community has played an important but ultimately limited role in Indonesia's power sector restructuring process. The World Bank had a long history of power sector project loans and policy analysis, and had encouraged the government to pursue corporatization as early as the 1980s. However, the increasingly scandalous IPP agreements led the World Bank to terminate lending to the sector in 1998, and its involvement in restructuring debates after 1998 has been limited. The IMF (at the World Bank's urging) included power sector reform as one of many structural conditionalities in its series

of bailout packages starting in 1998. The Asian Development Bank, through a $400 million loan, encouraged attention to public participation in the design of power sector restructuring, but many NGOs—not wanting to participate in activity funded by loans that would add to the country's debt burden—refused to participate. Finally, USAID has historically provided valuable advisors to the sector, but has been constrained by the emergence of IPP-related disputes.

The public benefits agenda received uneven attention in Indonesia's power sector reform debates. Concern about promoting equity in the context of power sector restructuring was widely shared among domestic and international, government and nongovernment constituencies, and was not controversial. The white paper produced in August 1998 included specific provisions to continue subsidies to poor regions and poor consumers, and dealt with labor issues. However, the relationship between electricity tariff increases and the public interest continues to be controversial.

Environmental implications of power sector restructuring received very little attention in the design process. A handful of NGOs have promoted more attention to efficiency and renewables, but environment was not a first-order consideration of any of the principals in the process. Improved governance has been a general theme of restructuring discussions, but the preponderance of attention has been focused on increasing financial transparency in the sector. Very little detailed consideration has been given to the structure and functioning of an independent regulatory body, which will play a critical role in determining who benefits from Indonesia's electricity sector reforms.

The Indonesia case demonstrates the challenges of incorporating public benefits into the electricity sector reform agenda. On the one hand, there is a sense of urgency among proponents of reform, who foresee power shortages constraining economic growth and poverty reduction if tariff increases and restructuring are not put into place to attract private investment in new generation capacity. On the other hand, measures to promote the public interest through reform—including social equity, environmental sustainability, and good governance—are quite complex, and require significant investment in analysis and public outreach to support informed public debate. Reconciling the need to move quickly and to safeguard the public interest will continue to be a challenge as the reform process moves forward.

NOTES

1. Most of the research for this chapter was undertaken in mid-2000, and was updated to reflect developments as of early 2002.

2. Frances Seymour would like to acknowledge research assistance provided by John Coyle.

3. Agus Sari is the Executive Director of Pelangi, a Jakarta-based NGO whose current and former staff have been active in the Working Group on Power Sector Restructuring. He would like to acknowledge the contributions of Ophelia Cowell, Rizka Elyza, Barbara Haya, Muhammad Suhud, Nasrullah Salim, and Fabby Tumiwa.

4. This chapter draws on a background paper for this study prepared by Agus Sari, 2001, "Power Sector Restructuring and Public Benefits, Who Cares?" which is available at http://www.wri.org/governance/iffepowercases.html.

5. For example, Presidential Decree No. 49 of 1991 provided tax breaks to geothermal generating companies to compensate risk and attract investment. Ministry of Mining and Energy Regulation No. 996.K/43/MPE/1999 prioritized PLN power purchases from small-scale generators of geothermal and other renewable energy sources above those based on fossil fuels.

6. For a description of how Indonesia was the "jewel in the Bank's operational crown" in that era, see Kapur, Lewis, and Webb (1997).

7. Interview with World Bank official, June 21, 2000. All interviews for this chapter were conducted on a not-for-attribution basis. Consequently, interviewees are identified only by their institutional affiliation.

8. Interview with World Bank official, June 21, 2000.

9. A World Bank official described the relative heft of the competing studies as "three big elephants versus a little mule" (Interview, June 21, 2000). Habibie continued to promote nuclear power into the mid-1990s.

10. Interview with World Bank official, July 10, 2000.

11. Interview with World Bank official, July 10, 2000.

12. Ironically, even as World Bank officials in Jakarta were trying to discourage the frenzy of PPAs, the World Bank Group's Multilateral Investment and Guarantee Agency (MIGA) was helping to facilitate them, providing a $15 million guarantee for U.S.-based Enron Corporation's participation in a power plant project. When the Indonesian government decided to suspend the project in 2000, Enron demanded compensation, resulting in a pay-out by MIGA (FOE, 2001).

13. Interview with World Bank official, July 10, 2000.

14. Interview with World Bank official, July 10, 2000.

15. Interview with ADB official, July 13, 2000.

16. This section draws significantly on material from "Power Sector Restructuring and Public Benefits, Who Cares?" by Agus Sari, 2001.

17. "Power Sector Regulatory Reforms," Coopers and Lybrand, (1996), not available to the authors. While donor agency officials point to this study as seminal to the subsequent reform agenda, Indonesian government officials downplayed its significance, saying "it was just a study" (Interview with staff of the Ministry of Energy and Mineral Resources, April 3, 2002).

18. Interview with ADB official, July 13, 2000.

19. Interview with international consultant, July 13, 2000.

20. Interview with ADB official, July 13, 2000.

21. Interview with World Bank official, July 10, 2000.

22. Interview with USAID officials, July 12, 2000.

23. Interview with USAID officials, July 12, 2000.

24. Interview with USAID officials, July 12, 2000.

25. Indeed, a recent study of electricity consumption, "Energy Outlook & Statistics: Indonesia 2000" projected that consumption would reach 281 terawatt-hours by 2020. Assuming that power is provided by 600 megawatt steam power plants, the capital investment needed would be approximately $90 billion, or $47 billion if met by combined cycle power plants.

26. Interview with World Bank official, November 15, 2000.

27. Interview with ADB official, July 13, 2000. A new Oil and Gas Law was eventually passed by the Indonesian Parliament in October 2001 (*Oil and Gas Journal*

Online, 2001), and hearings on the new Electricity Law began a month later.

28. Interview with international consultant, July 13, 2000.

29. Interview with USAID officials, July 12, 2000.

30. Interview with ADB official, July 13, 2000.

31. See, for example, Haugland, et. al. (1997), Center for Energy Research (2001), Tempo Interactif "YLKI Maklumi Kenaikan Harga BBM dan Listrik" (2002).

32. See, for example, *Tempo Interactif* "PRD dan A2PS Demo Tolak Kenaikan Harga BBM, Listrik, dan Telepon" (2002), Nusantara "Mahasiswa Nyaris Bentrok dengan Petugas Mahasiswa Domo Tolak Kenaikan Harga BBM" (2002).

33. At a public hearing on the Electricity Law in November 2001, NGOs expressed concern about the role of the MDBs in pushing the restructuring agenda. (Personal communication, Rizka Eliza, April 8, 2002). In April 2002, the NGO Working Group on Power Sector Restructuring published two reports that articulated this concern: *Listrik yang Menyengat Rakyat: Menggugat Peranan Bank-Bank Pembangunan Multilateral* [Electricity that Stings the People: Criticizing the Roles of Multilateral Development Banks] by Fabby Tumiwa, and *RUU Ketenagalistrikan Dalam Perspektif ORNOP* [The Electricity Law from the Perspective of Non-Governmental Organizations] by Tubagus H. Karbyanto.

34. Interviews with NGO leaders, July 11 and 13, 2000.

35. Interview with World Bank staff, June 21, 2000.

36. A study cited in the World Bank's (1994) report, *Indonesia Environment and Development*, referenced a 1992 study that conservatively estimated that 486 megawatts could be saved from a demand side management program.

37. The World Bank's more recent study on the environment in Indonesia (World Bank, 2001a) focused on management of terrestrial resources, and did not treat environmental issues in the electricity sector.

38. Chapter IX, Article 27 states "Any electric power business activity shall be obligated to conform with the required conditions in the prevailing laws and regulations on the environment."

39. The credibility of the Working Group was challenged in mid-2001, when it was revealed that a member of the Working Group representing the Indonesian Consumers Foundation was also a founder of an organization that received a six billion Rupiah contract from PLN. The contract was to produce materials to generate public support for electricity tariff increases, leading one Member of Parliament to observe that it gave the impression of PLN using the Foundation to ease public acceptance of the rate hikes (*Media Indonesia*, July 2001). While denying the existence of a conflict of interest, the individual resigned from his position at the Foundation and hence the Working Group.

REFERENCES

Asian Development Bank. 1999a. "ADB Loans a Big Step Towards Reforming Indonesia's Power Sector." Asian Development Bank Press Release (March 23). Online at: http://www.adb.org/Documents/News/1999/nr1999015.asp (May 31, 2002).

———. 1999b. *Report and Recommendation of the President to the Board of Directors on Proposed Loans to the Republic of Indonesia for the Power Sector Restructuring Program. (RRP: INO 31604).* March 1999. Manila: Asian Development Bank.

———. 2001. "A Fact Sheet: Indonesia and the ADB." Online at: www.adb.org/Documents/Fact_Sheets/INO.asp?p=ctryino (October 22).

Bisnis Indonesia. 2000a. "Kadin: Kenaikan TDL dan BBM Pukul Industri Nonekspor." [The Chamber of Commerce: The Increases in Electricity and Fuel Tariff Hit Nonexportable Industries]. (February 23).

Bisnis Indonesia. 2000b. "Industri Hotel Cemaskan Tarif Listrik" [The Hotel Industry Worried About Electricity Tariff]. *Bisnis Indonesia* (March 4).

Center for Research on Energy. 2001. *Kajian Harga Listrik Yang Rasional.* Jakarta.

Fried, Stephanie and Titi Soentoro. 2000. "Export Credit Agency Finance in Indonesia." Unpublished manuscript. Washington D.C.: Environmental Defense.

Government of Indonesia (GOI). 1998a. *Power Sector Restructuring Policy.* Jakarta: Ministry of Mines and Energy.

———. 1998b. Letter of Intent to IMF (November 13).

———. 1999a. Letter of Intent to IMF (March 16).

———. 1999b. Letter of Intent to IMF (May 14).

———. 2000a. Letter of Intent to IMF (January 20).

———. 2000b. Letter of Intent to IMF (May 17).

———. 2000c. "Konsep Akhir Rencana Umum Ketenagalistrikan Nasional" [Final Concept of the Master Plan for the National Power Plan]. Jakarta: Ministry of Energy and Natural Resources.

Harvest International. 2000. "Indonesia Agrees to Pay OPIC's Claim of $260 Million." Online at: www.harvest-international.com/perspec/jun2k1ef_1.htm (June).

Haugland, Torleif, Kjetil Ingeberg, and Kjell Roland. 1997. "Price Reforms in the Power Sector." *Energy Policy* 25(13).

Kapur, Devesh, John P. Lewis, and Richard Webb (eds). 1997. *The World Bank: Its First Half Century.* Washington, D.C.: Brookings Institution Press.

Karyanto, Tubagus H. 2002. *RUU Ketenagalistrikan Dalam Perspektif ORNOP.* Jakarta: NGO Working Group on Pwer Sector Restructuring.

Kristov, Lorenzo. 1995. "The Price of Electricity in Indonesia." *Bulletin of Indonesian Economic Studies,* 31(3). Canberra: Australia National University.

Media Indonesia. 2000. "DPR Akan Tolak Usulan Kenaikan TDL Jika Hasil Audit PLN Tak Diumumkan" [Parliament Will Reject the Tariff Increase Proposal if the Audit on PLN Not Made Public]. (February 9).

Media Indonesia. 2001. "Agus Pambagio Harus Mundur dari YLKI," [Agus Pambagio Forced to Pull Out of YLKI]. (July 9).

Motoyama, Hisako, and Nurina Widango. 1999. *Power Restructuring in Indonesia: A Preliminary Study for Advocacy Purposes.* Washington D.C.: Bank Information Center.

Norplan A/S. 1993. *Institutional Framework and Regulation of the Power Sector in Indonesia.* Washington D.C.: World Bank.

Oil and Gas Journal Online. 2000. "Indonesian Legislators End Pertamina Monopoly With New Law." Online at: http://ogj.pennet.com (October 24).

Pangsetu, Mari. 1996. *Economic Reform, Deregulation, and Privatization: The Indonesian Experience.* Jakarta: Center for Strategic and International Studies.

Pape, H. 1999. "Captive Power in Indonesia: Development in the Period 1980-1997." Paper presented at the half-day Seminar on Captive Power in Indonesia: Development, Current Status and Future Role. Jakarta: Indonesia (July).

Pelangi Indonesia. 2000. "Power Sector Restructuring Program." Unpublished manuscript. Submitted to the World Resources Institute.

Perusahaan Listrik Nasional (PLN). 1998. *Statistik PLN* [PLN Statistics]. Jakarta.

Perusahaan Listrik Nasional (PLN). 2000. *Empowering the Indonesian Villages.* Jakarta.

Rakyat Merdeka. 2000. "Kuntoro Didemo Ativis Gerakh" [The 'Movement Against Price Increase' Activists Staged

Demonstration Against (PLN President Director) Kuntoro]. (March 25).

Republika. 2000. "KAI Protes Kenaikan Tarif Oleh PLN" [Indonesian Railway Company Protests the Tariff Increase by PLN]. (March 2).

Sari, Agus. "Power Sector Restructuring and Public Benefits: Who Cares?" Unpublished manuscript. 2001.

Suara Karya. 2000. "Kenaikan TDL Jangan Pakai Asumsi Rasional" [Tariff Increase Should Not Use Rational Assumptions]. (February 9).

Suara Pembaruan. 2000a. "Kenaikan TDL Mempengaruhi Daya Saing Produk Indonesia" [Tariff Increase Influence the Competitiveness of Indonesian Products]. (February 22).

Suara Pembaruan. 2000b. "Tim Tariff DPR RI Klarifikasi Naiknya Harga Listrik dan BBM" [Parliamentary Team on Tariff Clarified Increase for Electricity and Fuel]. (March 2).

Sudja, Nengah. 1993. "Power Pricing Structure for Commercial Viability of Various Types of Power Projects." Unpublished manusript. Jakarta: Perusahaan Listrik Nasional.

Supplementary State Gazette of the Republic of Indonesia. 2000. "Draft Elucidation for Law of the Republic of Indonesia Concerning Electric Power." Jakarta: Government of Indonesia.

Sustainable and Peaceful Energy Network Asia (SPENA) Newsletter. 2001. Volume 3, Number 2, December 2001.

Tempo Interactif. 2002a. "YLK Maklumi Kenaikan Harga BBM dan Listrik" [The Indonesian Consumers Foundation Understands the Increased Tariffs of Fuel and Electricity]. (January 2).

Tempo Interactif. 2002b. "PRD dan A2PS Demo Tolak Kenaikan Harga BBM, Listrik dan Telepon" [PRD and A2PS Demonstrate Against the Increase of Fuel, Electricity, and Telephone Tariffs]. (January 7).

Tumiwa, Fabby. 2002. *Listrik yang Menyengat Rakyat: Menggugat Peranan Bank-Bank Pembangunan Multilateral.* Jakarta: NGO Working Group on Power Sector Restructuring.

Watson, Peter. 2001. Presentation to United States-Indonesia Society. Notes online at: http://www.usindo.org/Briefs/Peter%20Watson.htm (September 6).

World Bank. 1989. "Indonesia Power Sector Institutional Development Review." Washington D.C.

World Bank. 1994. *Indonesia Environment and Development: Challenges for the Future.* Washington, D.C.

World Bank. 1995. "Country Economic Memorandum for Indonesia." Report #14006-IND. Washington, D.C.

World Bank. 1996. "Implementation Completion Report: Indonesia Paiton Thermal Power Project." Loan 3098-IND Washington, D.C.: World Bank.

World Bank. 1997. "World Bank Approves Global Environmental Facility Trust Fund Grant." *World Bank Press Release* (June 25).

World Bank. 2000. *Indonesia Oil and Gas Study.* Washington, D.C.: World Bank Energy and Mining Sector Unit.

World Bank. 2001a. *Indonesia: Environment and Natural Resource Management in a Time of Transition.* Washington, D.C.

World Bank. 2001b. *Poverty Reduction in Indonesia: Constructing a New Strategy.* Washington, D.C.

6

BULGARIA

SUPPLY-LED VERSUS EFFICIENCY-LED ELECTRICITY REFORM

Dimitar Doukov[1]
Navroz Dubash
Elena Petkova

INTRODUCTION

Electricity sector reforms in Bulgaria have taken place under the shadow of extreme economic crisis and under the watchful eye of the International Monetary Fund (IMF). This crisis context is similar to that of several other studies in this study: a history of inefficient operation, flawed price signals, and the potential social costs of price reform. But the Bulgaria case, and that of Central and Eastern Europe in general, also differs from reforms in developing countries in several ways. Most significantly, access to electricity in Bulgaria is more or less assured, and the country has surplus generation capacity. In addition, Bulgaria bears significant international environmental commitments that raise the profile of environmental concerns. With commitments under the Kyoto Protocol, along with a dramatically enhanced environmental profile required for accession to the European Union (EU), environmental concerns presumably would rank high on an electricity sector reform agenda. For a profile of the electricity sector in Bulgaria, see Box 6.1.

Despite these circumstances, reforms in the electricity sector have been dominated by short-run economic concerns. Where social factors have played a role, it has largely been in postponing socially difficult decisions rather than finding lasting solutions. In spite of external pressures, reforms have proceeded without explicit attention to opportunities for environmental gains. The failure to anticipate either social or environmental concerns is arguably

BOX 6.1	PROFILE OF THE ELECTRICITY SECTOR IN BULGARIA

Population (2001)[1] : 8.1 million

Percentage of households with access to electricity[2]:
  Rural: 100% *Urban: 99.9%*

Installed electricity generation capacity (1999)[3]
Total: 12 gigawatts (0.37% of total world capacity)
  Thermal: 58%
  Hydro: 17%
  Nuclear: 33%
  Geothermal and Other: 0

CO_2 emissions from electricity and heat as a share of national emissions (1999)[4] : 56%

Notes

1. World Resources Institute. 2000. *People and Ecosystems: The Fraying Web of Life*. Washington, D.C.: World Resources Institute.

2. Alan Townsend. 2000. "Energy access, energy demands, and the information deficit." *Energy Services for the World's Poor*.

3. www.eia.doe.gov/pub/international/ieapdf/t06_04.pdf (February 6, 2002).

4. Computed by WRI using International Energy Agency (IEA) data, 2001. *CO2 Emissions from Fossil Fuel Combustion*. Paris: OECD.

rooted in the conflicting incentives faced by various government actors, the delayed attention given to these issues by donor agencies, and the lack of an active, vocal, and capable constituency to advance attention to public benefits. Fortunately, there are hopeful signs of change; both a new government elected in 2001 and the World Bank have introduced initiatives that pay far greater attention to the social and environmental dimensions of reform.

Macroeconomic, Social, and Political Crises

Electricity reform in Bulgaria took place during a decade of economic and political change, when the nation began the difficult transition from a socialist to a market-led economy and society. This transition has been marked by several crises.

In 1991, the government initiated an economic stabilization program based on restrictive monetary and fiscal policy.[2] The resulting budgetary restrictions led to social unrest in the early 1990s, forcing somewhat less restrictive policies. By 1994, a currency crisis led to a near doubling of the exchange rate in just one year and a steep increase in inflation to a 90-percent average annual rate. Between 1995 and 1997, a precipitous drop in hard currency reserves, burgeoning debt payments (Todorov, 2001), and the failure to negotiate a successful agreement with the IMF all combined to undermine public confidence in the financial system. In 1996, private citizens responded with a run on bank deposits.

The economic crisis, and the imposition of austerity measures to tame it, took a terrible social toll. Between the early 1990s and the height of the crisis in 1996, the average wage had fallen by about 80 percent, the average pension by 84 percent, and the "guaranteed minimum income" by 70 percent.[3] Due in part to the insolvency of state enterprises, unemployment levels soared, reaching 15 percent in 1997 (National Statistical Institute, 1998). As employment dropped, so did payments into the country's social insurance system. By mid-2000, when energy sector reforms were to begin, the country faced a fragile economy recently emerged from deep crisis, a society losing patience with a decade of declining living standards, and a government that sought to balance responsible financial policies and commitments to donors, against the demands of a weary population tired of austerity.

The Pre-Reform Sector

As in many other parts of Central and Eastern Europe, the pre-1989 electricity sector in Bulgaria was shaped by several strategic concerns. At minimum, policymakers in the sector considered it important that the country was fully electrified and that regular and reliable supplies were available to power the economy. In addition, energy security has been and continues to be a major concern. Bulgaria is relatively energy poor. It has traditionally relied on imports, primarily from Russia, for between 65 and 70 percent of its fuels (National Statistical Institute, 1993). While importing primary fuel, Bulgaria has been an exporter of electricity to neighboring countries. By the late 1990s, the country generated surpluses for export on the order of 2,000 megawatts, which provided much-needed revenues for the state budget. Electricity exports also serve political ends, which have changed over the years. Over the last five years, exports to neighboring Turkey have served to strengthen political and economic ties with this potentially large customer for Bulgarian power. Finally, the energy sector (particularly when the coal sector is included) is a major employer, accounting for 4.5 percent of all jobs available in Bulgaria in 1999 (National Statistical Institute, 2000).

As a result of these concerns, the pre-reform sector was integrated and control was centralized under a government agency. The Energy Committee (renamed on many occasions between 1989 and 2001) was responsible for policy development for the coal and natural gas sectors and for electricity generation, transmission, and distribution. This body, later reconstituted as the State Agency for Energy and Energy Resources (SAEER), had primary responsibility for restructuring energy institutions in the country. In December 2001, it was renamed yet again as the Ministry of Energy and Energy Resources. In 1997, a National Energy Efficiency Agency was also created, later transformed into a State Energy Efficiency Agency.

The Energy Committee had policy responsibility, but the responsibility for managing and operating the generation, transmission, and distribution of electricity was almost exclusively handled by the state-owned National Electricity Company (Natsionalna Elektricheska Kompania - NEK). This two-tier arrangement, which existed with minor modifications until about 1999, led to considerable lack of transparency in financial management. The two institutions shared subsidies and revenues in a manner that undermined financial responsibility. For example, revenues were allocated through an extra-budgetary Energy Resource Fund, which was managed by a board of directors and chaired by the president of the Committee of Energy. The electricity, coal mining, and district heating sectors were linked by the opaque shuffling of finances through the Energy Resources Fund. With little knowledge of the true cost of inputs, and therefore of true returns in the sector, there was little scope for prudent financial management. The financial implications of these arrangements were serious, since the Energy Resource Fund was considerable. The expected income of the fund was BGL 268 billion, ($152 million), or 6.42 percent of the State Budget for 1998 (Alexandrova, 1997).

An additional problem in the sector was the poor condition of the nation's power infrastructure. Prior to reforms, NEK controlled 90 percent of generation, with the rest owned by district heating companies. Many of the coal-powered plants, which together with other thermal sources account for about 48 percent of generation, were built in the 1950s and 1960s. The Kozloduy nuclear power plant, which provides about 44 percent of all electricity generated in

Bulgaria, is an old design and among the most risky in Europe. Hydropower plants provide the remainder 7 percent of generation (National Electric Company, 2001). Due to age and poor maintenance, the system as a whole is highly inefficient. Between 30 and 35 percent of power generated is lost during transmission and distribution, in part due to theft. With high reliance on coal, the sector contributes significantly to local air pollution and to emissions of greenhouse gases. Energy production and transformation activities actually are the single most important contributor to greenhouse gases, and account for about 65-70 percent of the country's total carbon dioxide emissions (Republic of Bulgaria Council of Ministers, 2000a).

From the early 1990s, donor agencies played a significant role in attempting to reform the sector, largely through a series of studies and programs of technical assistance. These included efforts at load forecasts for the sector; a least-cost generation and transmission expansion program; a tariff study; a study on approaches to reorganization of the sector; and a feasibility study for retrofitting and rehabilitation of thermal power projects (World Bank 1992a; 1992b; 1993; Center for the Study of Democracy, 1995). A range of agencies were engaged in these efforts, including the World Bank, the Commission of the European Community, United States Agency for International Development (USAID), United States Environmental Protection Agency (USEPA), and United States Trade and Development Program (USTDP). A World Bank summary study recommended increasing efficiency through more realistic pricing policies and energy saving investments; expanding domestic production and reducing imports; and improving safety and the environmental performance of the sector (World Bank, 1992b; 1993).

In addition to studies on the energy sector, donor agencies also conducted studies of Bulgaria's environmental performance. A March 1992 joint World Bank, USAID, and USEPA study on an environment strategy for Bulgaria concluded that the poor state of the environment was rooted in the flawed economic and management policies of the transition period

(World Bank, 1992a). It suggested that market reforms, particularly privatization, would bring tangible environmental improvements. The report assumed that environmental improvements would be an indirect benefit of market reform, and did not emphasize the need for environmental protection and consideration apart from market liberalization. A subsequent December 1994 World Bank study did find marked improvements in the environmental situation. Distressingly, however, these improvements were largely a result of declining economic activity (World Bank, 1994).

Environmental concerns were an input into the design of a World Bank Bulgaria Energy Project initiated in 1993. The project aimed at improving the operating efficiency of the system; introducing commercial practices at NEK; realigning, improving, and depoliticizing the tariff structure; reducing the need for high-cost imports of electricity; and increasing safety in the system (World Bank, 1993). These were to be accomplished through two technical components and a third component based on technical assistance and directed at institutional reform. Drawing from the environment strategy study, the energy project aimed to achieve these goals by increasing energy conservation and efficiency of use. This approach foreshadowed subsequent reform efforts.

However, implementation was limited for political reasons. First, the government changed hands in 1993, leading to a dismissal of the management team of the Committee of Energy and NEK. While the new team made progress on one of the technical components of the project, the real casualty was the institutional reform component. The government's overall policy was to maintain the public monopolies in critical sectors such as electricity generation. This policy stalled reform in the electricity sector. The donors promoted commercialization in the sector as a first step in breaking the monopoly—to be followed by a separation of power generation, transmission, and distribution into separate entities.[4] But the new socialist government contended that the structure of the sector, built around the premise of a centrally planned economy, precluded commercialization. The

government's main anti-reform argument was technical. They argued that each power facility was built to support a specific load of the system, and the opportunity for real competition between them was very limited.[5]

While the government rejected a reform trajectory based on eventual private ownership of the sector, it failed to formulate an alternative vision and did not undertake the financial and management reforms needed to put the sector on a sound footing.

In addition to donor engagement, Bulgaria also faced external pressures pertaining to international political obligations. Bulgaria is a signatory to the UN Framework Convention on Climate Change. As an Annex 1 country under the Kyoto Protocol, it will have to reduce its emissions by 8 percent from a 1988 baseline level during the period 2008–12. However, as with other countries in the region, aggregate greenhouse gas (GHG) emissions have declined precipitously due to an economic slowdown, and by 1997 were 40 percent below 1988 levels (Council of Ministers, 2000a). Nonetheless, there are good reasons to continue to pay attention to these emission levels. First, emissions are likely to grow as economic growth picks up, and second, emission reductions offer potentially valuable financial opportunities to Bulgaria in an emerging market for greenhouse gas reductions.

Along with international climate obligations, Bulgaria also has obligations as a candidate for EU membership. These commitments are covered in a series of agreements that concern nuclear safety, integration of energy markets and promotion of competition, and improvement of energy efficiency and reduction of environmental harm from the sector (Energy Charter Secretariat, 1996). In particular, the EU electricity directive requires all members to introduce competition in the sector, although it leaves open whether there must be a complementary shift in ownership in the sector.[6]

In short, with the move to a market-based system as part of a larger political and economic transition in Bulgaria, several prior assumptions remained open to challenge. A market-based approach called into question the state's ability to manipulate the sector based on energy security concerns. A more transparent fiscal system, which could send signals to market participants, potentially undermined the financial web that supported the sector. And with pressing international environmental obligations, the environmental profile of the sector stood to be altered by a reform approach.

The Reform Trigger

The immediate trigger for electricity sector reforms was a dramatic escalation in the economic crisis in the mid-1990s. Inflation grew by 552 percent between January and March 1997 (National Statistical Institute, 1998). By late 1996, GDP was only 58 percent of its 1990 levels (National Statistical Institute, 1994; 1999). Faced with this unsustainable position, the socialist government resigned in February 1997, creating a political crisis.

The crisis showed signs of abating in April 1997, when a new government with a majority in Parliament came to power on a pro-reform platform. This administration negotiated a stabilization agreement with the IMF in May 1997 that was based on two main conditions. First, it committed to a program of fiscal discipline and austerity, with a stable macroeconomic policy to be anchored by a currency board. Moreover, the agreement required liquidation or privatization of 50 percent of the long-term material assets under control of branch ministries or municipalities. Efforts at stabilization did have positive short-term results—GDP growth went from negative to positive rates, inflation came under control, and the budget deficit shrank between 1996 and 1999. At the same time, the price of stability was reduced national control over decisionmaking across a broad range of economic and social policies. In particular, state expenditures for social policy had to be negotiated directly with the IMF. Another important condition of the IMF package was reform of the energy sector. Hence, the financial crisis and IMF conditionality as part of a stabilization package was a direct trigger for reforms in the energy sector.

The Reform Design

Under the proposed structure, the post-reform sector in Bulgaria would operate along the lines of a "single-buyer" model. As a first step, the design required that the property and assets of NEK be clearly delineated among generation, transmission, and distribution units, including a clear accounting division of existing credit arrangements. Based on this allocation of property rights, several generation units, particularly the hydro and thermal units, were to be offered for privatization. Two of the Kozloduy nuclear plant's six units were scheduled to be decommissioned, while others were to be upgraded to meet safety standards. The nuclear plants were not scheduled for privatization and would remain under the control of the government.[7] With regard to distribution, the three-year plan of action agreed to with the IMF envisaged that distribution companies would be legally separated from NEK by the end of 1999. This was eventually accomplished in 2000. After this unbundling of NEK, based on a clear accounting separation, a publicly owned transmission company remained, with additional responsibility for planning, investment, dispatch, and power trade functions.

Under this approach, the new public National Electricity Transmission Company would purchase electricity from the independent producers and sell it to the distribution companies, which would then sell it to consumers. In addition, the transmittion company would be able to sell directly to large consumers connected to the high-voltage transmission network, and independent producers would also be able to sell electricity directly to "privileged consumers," bypassing the transmission company. *(See Figure 6.1.)*

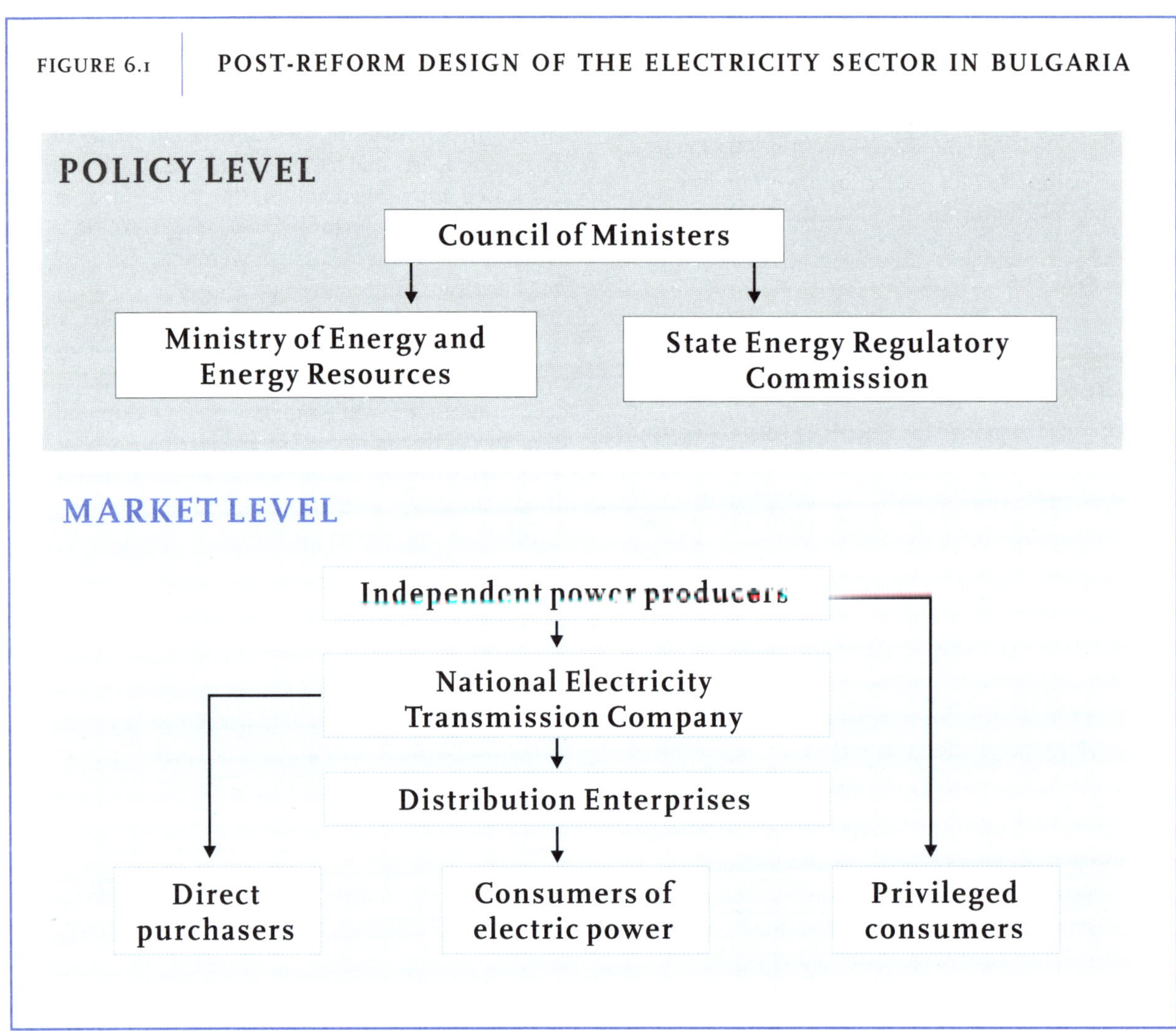

In order for this approach to work, the IMF further required that prices in the sector increase to cover the anticipated investment costs and provide a profit to private investors. Thus, by July 2001, the price for households and industries was to be equalized at 4 cents per kilowatt-hour—a considerable increase for households from the then-prevailing rate of 3.5 cents per kilowatt-hour and below.

The defining moment in the reform process came in July 1999, when an "Energy and Energy Efficiency Act" was passed in Parliament, fulfilling a key condition of the IMF agreement. Soon thereafter, in September 1999, the State Energy and Energy Resources Agency (SAEER) replaced the earlier Committee on Energy. In addition, the State Commission on Energy Regulation was created as independent from the government regulatory body. It would be the Commission's task to make decisions on prices, and on the issuing of licenses and permits, and to do so in an independent fashion, free from political interests (Republic of Bulgaria National Assembly, 1999).

Ensuring financial accountability was central to the initial reform design, which was shaped largely by the short-term economic crisis in the sector. Over time, a range of different actors and interests within and outside the government gradually asserted their influence, which led to an implementation trajectory somewhat at odds with the original design. In 2001, the election of a new government initiated a reexamination of the reform approach and renewed efforts to work with donor agencies. This section provides a sketch of the various actors and their motivations, and concludes with an assessment of reform implementation efforts thus far. A chronology of significant events is provided in Box 6.2.

Governmental Actors

In the course of reforms, a range of governmental actors faced contradictory incentives that influenced, at different times, the pace and trajectory of reforms.

The extra-budgetary Energy Resource Fund provided an extremely useful political tool.

For the coalition government that came to power in May 1997, broad sentiments were in favor of reform. The government had been voted in on a platform of managing the fiscal and monetary crisis in the country. In the short term, this required not only reducing government expenditures by increasing prices and reducing subsidies, but more generally keeping the flow of IMF money coming by complying with IMF conditions, among them reform of the electricity sector. Moreover, the privatization of electricity generation facilities promised to bring in revenues. Finally, progress toward market-based reform brought political benefits in the form of progress toward integration with the EU.

These political motivations for the party in power were considerably undermined by other, quite

different motivations for politicians and bureaucrats with direct control over the resources in the energy sector. The extra-budgetary Energy Resource Fund provided an extremely useful political tool, free of parliamentary supervision. By allocating funds and subsidies, politicians could create or foster political allies, or defuse potentially dangerous situations. Since the reform measures were aimed at introducing financial transparency, reforms threatened to undermine a potent political tool.

Tariff increases carried enormous political dangers for the government in power.

In addition, while international donors would reward the government for progress on reforms, many of the same measures were sure to incite the wrath of the voting population. Tariff increases, a central component of reform efforts, carried enormous political dangers for the government in power. As a result, it consistently pressed the IMF to allow slower implementation of tariff increases and subsidy removal, leading to an ongoing tension in the reform process. In particular, the government aimed to manage this tension with targeted social assistance,[8] slower than planned price increases for households, and continued subsidies to the politically important district heating sector.

Moreover, several elements of the energy bureaucracy had a strong disincentive to slow or otherwise subvert reforms. With the implementation of reforms, bureaucrats had to deal with layoffs. Through manipulation of nontransparent funds, energy sector management stood to lose salaries that had been substantially insulated from the economic downturns that had damaged other sectors of the economy. In particular, the emergence of an independent agency charged with overall management of the sector threatened their interests.

In 1999, the government passed a "National Strategy for Development of Energy and Energy Efficiency until 2010." This strategy envisaged a

CHRONOLOGY OF ELECTRICITY SECTOR REFORM IN BULGARIA

October 1991	Law on Protection of the Environment.
November 1991	Establishment of the Committee of Energy and the National Electric Company (NEK).
December 1991	The European Energy Charter approved in The Hague.
May 1992	Law on Transformation and Privatization of State and Municipal Enterprises.
March 1995	Law on Ratification of UNFCCC.
September 1996	Creation of Ministry of Energy and Energy Resources (Closing down of the Committee of Energy).
May 1997	Three-Year Agreement with IMF creating conditions for long term financial stabilization and economic growth.
May 1997	Creation of Committee of Energy to replace the Ministry of Energy.
May 1997	Creation of National Agency for Energy Efficiency.
July 1997	Implementation of Currency Board in Bulgaria as a result of agreement with IMF.
August 1998	Approval of a Plan of Action for Restructuring, Abolishment of Subsidies, and Financial Rehabilitation of the Commercial Societies in the Energy Sector during the Period 1998–2000.
March 1999	Decision by the National Assembly on the National Strategy for Development of the Energy Sector and Energy Efficiency until 2010.
July 1999	Energy and Energy Efficiency Act.
September 1999	Decree 179 of the Council of Ministers on the Transformation of the Committee of Energy into State Agency for Energy and Energy Resources.
September 1999	Decree 180 of the Council of Ministers on the Transformation of the National Agency for Energy Efficiency into State Agency for Energy Efficiency.
April 2000	Decree 181 of the Council of Ministers on the creation of a State Commission on Energy Regulation.
August 2000	Restructuring of the National Electric Company and registration of independent electricity producers and distribution companies.
July 2001	Approval of a Strategy for Development of District Heating during the period 2000-2005.
February 2002	Election of a new Government of the National Movement Simeon II.
February 2002	Closing of the State and Energy Resources Agency and approval of the Statutes of the Ministry of Energy and Energy Resources.
March 2002	Transformation of the State Energy Efficiency Agency and approval of the Statutes of the Energy Efficiency Agency.
March 2002	Release of an Energy Strategy for the Republic of Bulgaria by the Ministry of Energy and Energy Resources.

considerable increase in new hydro, nuclear, and thermal supply capacity. This projection was based on the assumption that existing industrial capacity was being underutilized, and that demand for electricity by both industrial and household segments would increase by 5 percent a year as the economy returned to pre-1990 levels (Republic of Bulgaria Council of Ministers, 1998b). Moreover, the strategy assumed a considerable electricity shortage in the region. Consequently, a substantial motivation for new generation capacity was to develop Bulgaria as an electricity exporter. This policy had considerable implications for the reform process.

For two principal reasons, the evidence suggested that Bulgaria would not be well-positioned to turn a profit from electricity sales. First, with the exception of Turkey, other countries in the region had only a temporary shortage of electricity. This suggested that Bulgaria would not be able to command a premium price.[9] Second, low-price production depended on a stable supply of low-cost imported fuel. This situation could not be expected to continue as Bulgaria's fuel suppliers liberalized their own energy sectors. Nonetheless, elements within the electricity bureaucracy and the government as a whole saw a future for Bulgaria in the electricity markets, and were determined to forge ahead with this vision. Their main arguments were that exports would allow the country to retain substantial generation capacity until domestic demand increased, and that electricity was the only commodity that Bulgaria could competitively export, at least in the short term.

Energy sector management stood to lose salaries that had been substantially insulated from the economic downturns.

In 2002, there was a sharp change in direction. Following the election of a new government, the reconstituted Ministry of Energy and Energy Resources (formerly SAEER) drafted a new and considerably different "Energy Strategy of Bulgaria." This Strategy emphasized that Bulgarian energy had to be competitive on the Balkan market, placed a high priority on energy efficiency, and emphasized the need for a strong regulator as a precondition of establishing a market framework (Ministry of Energy and Energy Resources, 2002). The 2002 Strategy noted that the earlier 1999 Strategy took for granted growth in household consumption, and failed to apply a proactive approach to energy efficiency. The 2002 Strategy aimed to correct these shortcomings.

International Financial Agencies

With its central role in designing a macroeconomic stabilization program, the IMF was initially the most significant donor agency shaping the electricity sector in Bulgaria. The IMF advocated price increases; strong regulatory development to keep price-setting independent; dismantling of NEK's monopoly; privatization; and removal of subsidies. The IMF agenda was supported by other donors such as the World Bank (Tellam, 2000).

A substantial triumph for the IMF, and a positive outcome for transparency in the sector, was the abolishment of the Energy Resource Fund—the primary mechanism for nontransparent financial management. At the insistence of the IMF, the 1999 Law on the State Budget abolished the numerous existing extra-budgetary funds, and the money was remitted to the state budget (Republic of Bulgaria National Assembly, 1998). At the same time, the IMF's insistence on closing extra-budgetary funds also had unfortunate side effects. A Fund for Energy Efficiency Projects was dismantled, cutting off the only source of funding for energy efficiency in the country, with no provision made for pursuing these projects through more transparent means. Eliminating the extra-budgetary funds, however, did not eliminate the subsidies for which they were used—primarily for district utilities.

The government and the IMF have had two substantial differences in implementation of the reform process. First, the IMF was very clear that they foresaw unbundling of NEK only after a clear legal and regulatory framework for the sector had been put in place. [10] Based on accounts of reform elsewhere, this sequencing is necessary, and the condition imposed by the donor agencies is a sound one. Second, the IMF was strongly opposed to the government's decision to freeze electricity and district heating prices for households until the end of 2001. From the IMF's perspective, the freeze in prices risked reducing the "trust in the commitment of the Government to equalize the prices for households and industry; it will reduce NEK's profit and…will undermine the governmental plan for privatization of the generation and distribution components of NEK" (Kahkonen, 2000). From the government's point of view, price increases imposed a high social cost, one that also risked a political backlash.

Despite these concerns, the President of SAEER had already signed the orders for the separation of seven electricity distribution branches and four thermal power plants, together with the Kozloduy nuclear power plant. What was left of NEK after this unbundling was registered as a single-buyer transmission company. These changes occurred well before regulatory arrangements were satisfactorily in place. Hence, while reforms were certainly undertaken at the behest of the IMF, it had limited control over the subsequent implementation process.

During the initial stages of electricity reform, the World Bank was a junior partner to the IMF. More recently, it has taken the lead in engaging the government on sector reforms. The opening for the World Bank to re-enter the process came through an IMF condition that required the Government of Bulgaria to present a suitable law on energy sector reform to Parliament. The World Bank offered assistance in preparing this law—a process that began in October 1998. While the government proved initially reluctant to accept World Bank engagement, the Parliament proved more forthcom-

ing. [11] From a World Bank perspective, initial drafts of the law had considerable flaws, including a continued supply orientation, maintenance of NEK as a vertically integrated monopoly, and the continued subjugation of the Energy Commission to the Energy ministry. [12] In addition to the World Bank, USAID played an important role in providing legal assistance with drafting the law.

The World Bank has most forcefully articulated a concern that the Bulgarian government's efforts to build new capacity through Independent Power Producers (IPPs) casts a long shadow over reform. Together with the IMF, the World Bank has expressed fears that in the case of default, the government would be exposed to severe liabilities. In addition, the World Bank has argued that the projects are unnecessary, and questioned their financial viability.

These points have been made in the context of an "Energy–Environment Review" in Bulgaria, which was first reviewed by the government in November 2000 and subsequently published in February 2002. This is the first study to deal explicitly with the impact of reforms on the environmental profile in the sector. The report develops a scenario based on energy efficiency, which envisages, among other things, preferential dispatch of "green" power. Under this scenario, the life of nuclear units would not be extended, electricity exports would be limited, and the investment program of NEK would be reviewed and potentially halted. Based on this and other such analyses, the most incendiary conclusion of the paper is that Bulgaria does not need new generation capacity in the coming years, in complete contradiction to the government's own original 1999 Strategy document.

The IMF agreed with the World Bank that new power projects were undesirable. Yet, once it became clear that the government was proceeding with new deals, the IMF appeared to soften its disapproval. This accommodation may have been influenced by the near conclusion of its three-year agreement with the government, and the need to negotiate a new agreement. Thus, the government and its two major international financing institutions were sending contradictory and confusing messages.

The World Bank has also prepared the only study
that estimates the cost of meeting various environmental commitments. A draft of the study—"Environment Sector in Bulgaria: The Challenge of
Preparing for EU Accession"—was completed in
March 2000. The results are sobering. In the
electricity sector, the study estimates price increases
between 240 and 300 percent to cover necessary
environment-related investments.

Both these studies provided useful information and
analysis. However, they came late in the reform
process. By the time the studies were released, NEK
had already been unbundled, and the government
had committed to new generation capacity. Had such
convincing studies on alternatives to new capacity
and the social cost of reforms been available early in
the reform process, the information would have been
far more likely to shape both the weight given to
energy efficiency and the role of new capacity.

With the election of a new government in 2000, the
World Bank has taken an active role in preparation of
an ambitious $450 million "Programmatic Adjust-
ment Loan" (PAL) for the electricity sector.[13] This
loan will likely focus on tariff reform, regulatory
measures, privatization, and social protection. As the
World Bank has taken the lead in shaping donor
approaches to the sector, the IMF has increasingly
aligned its own recommendations with World Bank
prescriptions for the sector, reversing the dissonance
that characterized donor interaction with the government in the late 1990s.[14]

Transparency and the Scope for Participation in the Reform Process

The reform process in Bulgaria has been almost
entirely a government-led affair. Debate has been

conducted between government agencies and donor
institutions, primarily the IMF. Within the government, SAEER has played a dominant role. In the
latter stages of implementation, SAEER has been
making decisions independent of donor agency
views.

The government has published some information
on the reform program, but it has been incomplete
and allowed few opportunities for debate. For
example, it released a "Program for the Restructuring
of NEK," along with a calendar of achievements and
regulations for implementation of the Energy and
Energy Efficiency Law. However, prior to its enactment, the law was not available for public viewing or
comment.

Within civil society, the most vocal actors were the
trade unions in the energy sector. The unions stood
to suffer considerable losses as a result of the reform
process. In a letter to Parliament, the Prime Minister,
and the Deputy Prime Minister, these unions
critiqued the reform model being proposed by
SAEER, charging that several of the measures ran
the risk of bankrupting the sector, causing further
loss of jobs (*Pari Daily*, 2000). However, the unions
took no further steps; in particular, they made no
effort to combine forces with other concerned
citizens. That such alliances are possible is demonstrated by the example of South Korea. *(See Box 6.3.)*

Consumer groups and other NGOs with concerns
larger than those of sector employees have played a
far more muted role. Some energy efficiency and
environmental NGOs have provided technical
comments to the initial drafting stage of the law, but
have been little engaged in the implementation
process. From the point of view of consumer groups,
the rise in energy prices is of considerable concern.
Despite the significant social costs of reforms to
consumers, however, there has been little mobilization around this issue, in part because the government has postponed difficult decisions through a
price freeze. There are primarily two reasons for the
modest role NGOs and other public interest groups
have played in the reform process or in mobilizing
public opinion. First, few NGOs in Bulgaria focus on

the intersection of energy and environment, and none has the ability to support an informed public dialogue on such a complex issue. Second, little information was available to the public. There was no mechanism for public input on energy pricing, social outcomes and necessary safety nets; on reform objectives and the process to achieve them; on national goals for development of the energy sector; and on other major decisions with significant environmental and social outcomes.

Initial Outcomes

The reform process has proceeded more slowly than planned. The government's understandable concern about raising prices too far too fast, and the contradictory interests within various government agencies, including a concern that reforms be made irrevocable before elections, have shaped the halting progress of reforms. Since the initial agreement, this unsteady progress has caused significant tension

benefits concerns such as price hikes, cutting of public benefit programs, and difficulties in regulating powerful private companies post-privatization. Environmentalists have expressed sympathy for broader initiatives against the undermining of national authority by global forces, while noting that the environmental case for the electricity sector argues for market discipline. They also cautioned against a rosy view—then prevalent among labor groups—of the past achievements of state-owned corporations like KEPCO.

Ultimately, agreement on a proposal for restructuring sponsored by environmentalists failed to win agreement from labor unions over the issue of privatizing nuclear power. Environmentalists supported it and labor groups were strongly opposed. While the CPD-led process did not lead to a common platform, the experience led to each adopting positions that were more sensitive to the concerns of the other. Unions now stress the need to democratize the governance of the sector. Environmental groups acknowledge the danger that reform could create ungovernable private oligarchs, and have focused their efforts on decentralized power options and incorporation of environmental standards. A lesson from the Korean experience with a red-green alliance is that while the interests of each group are considerably opposed, there is

certainly space to forge coalitions. At minimum, the effort has resulted in greater mutual appreciation of the broader challenge of creating a responsible and beneficial electricity sector.

Sources:

Cho, Sung-Bong. 2000. *Restructuring and Privatization of Electric Power Industry in Korea.* Korea Energy Economics Institute, Euwang, Korea. (in Korean).

Jang, Youngsik. 2000. "Restructuring and Privatization Program of Korea Electric Power Corporation." A paper presented at the Center for Energy and Environmental Policy (CEEP) Colloquium Series, Newark, Delaware.

Kim, Yun-Ja. 2000. "Problems of the Privatization of the Infrastructure Industry and an Alternative Proposal." Presented at a Conference titled Privatization: What are the Problems, FKTU, Seoul, Korea.

Lee, Pil-Ryul. 2000. "Direction of Electricity Restructuring Viewed from the Standpoint of Energy Transition." Unpublished manuscript. (in Korean).

Oh, Kun-Ho. 2001. "For Further Development of an Argument Against the Privatization of the Network Industry." Online at www.kctu.org. (in Korean).

Soh, Ju-Won. 1998. "Problems of Electricity Monopoly and Basic Principles for Electricity Restructuring." Presented at the Conference on the Direction of Electricity Restructuring for Environmental Protection, Local Participation, and Energy Self-reliance, Korea Federation of Environmental Movements, Seoul, Korea. (in Korean).

with the IMF. Despite its leverage, the IMF has ultimately accepted the government's decisions. Two specific interim outcomes threaten to have considerably negative long-term impacts on reforms.

First, NEK was dismantled well before institutional and regulatory capacity was created and put in place in the sector. In part, this was likely because the government sought to make reforms irreversible before the elections in 2001. While this reverse

sequencing was in direct opposition to donor requirements, the donor agencies ultimately chose not to question the decision. However, the lack of a regulatory framework created an ambiguous environment for operation of the newly unbundled entities, and did not provide the clear signals necessary for them to operate on a commercialized basis. If privatization proceeds under these circumstances, there is considerable potential for manipulation of the privatization process for both financial and political gain. In recognition of these problems, the Ministry of Energy and Energy Resources' March 2002 Strategy document explicitly recognizes the problem of unbundling before undertaking regulatory reform.

Second, the momentum toward new thermal power capacity—at Maritsa-Iztok 1 and 3—prior to the restructuring of NEK is a cause for concern. The contract proposed would oblige NEK to purchase electricity on a "take or pay" basis, forcing it to bear the risk of a downturn in demand for electricity. The price at which this electricity will be purchased is kept confidential. That the contract is for a 15-year period contradicts the spirit of the reforms, which are intended to shift the sector toward a more competitive structure. The government currently anticipates that the electricity will be exported to Turkey. However, based on estimates of generation costs and the likely price of power exports once the market is fully liberalized and linked to the European grid, there is a high likelihood that the electricity will be sold at a considerable loss. If this does prove to be the case, Bulgarian consumers, who are already struggling under a heavy price burden, will effectively subsidize cheap exports to Turkey. The reform model squarely puts the burden on the small consumer, since "privileged" consumers such as large industrial facilities can buy directly from producers at competitive prices, depriving the distribution companies of their most lucrative customers. Consequently, the distribution network may have to pass on the costs of more expensive supply to small consumers and households. In brief, the public sector and ultimately the small consumers in Bulgaria bear the project risks, while private sector partners and the importing country stand to benefit.

ENVIRONMENTAL AND SOCIAL CONCERNS

Until 2001, there has been little evidence in the reform process of attention to either environmental or social outcomes. This is despite substantial international pressures on Bulgaria to meet environmental targets, and despite palpable evidence that the reform process is extracting a considerable social cost on the most vulnerable.

The design and implementation of reforms has not led to the emergence of any domestic champions for environmental concerns. The plans for restructuring and privatization were drawn up prior to the World Bank studies on energy and environment completed in 2000. Consequently, reform design was drawn from the decade-old World Bank studies, which concluded that market reforms would automatically result in environmental improvements. The possibility that reform processes might undermine efforts at environmental improvements was not considered or mitigated. For example, the possibility that creating seven separate distribution companies could substantially complicate implementation of end-use efficiency programs has not been considered. Nor was any systematic treatment given to the scope for pursuit of renewable energy options in the reform process.

If fully implemented, the government's original 1999 Strategy document, which focused on supply expansion, would have likely had negative environmental implications. The 1999 Strategy was based on the assumption that the expected economic recovery would lead to demand comparable to pre-1989 levels, and on a vision of Bulgaria as a major energy exporter. This approach placed limited effort on supply and demand efficiency. Currently, even the existing funds for energy efficiency projects have been lost as part of the program of fiscal transparency. As examples from other Central and Eastern European countries indicate, sector and macroeconomic reform provide incentives for energy efficiency in the privatized industries, but they are not equally effective in the public sector or with households. Special instruments such as earmarked funds and soft financing are needed to introduce energy efficiency measures (Regional Environmental Center

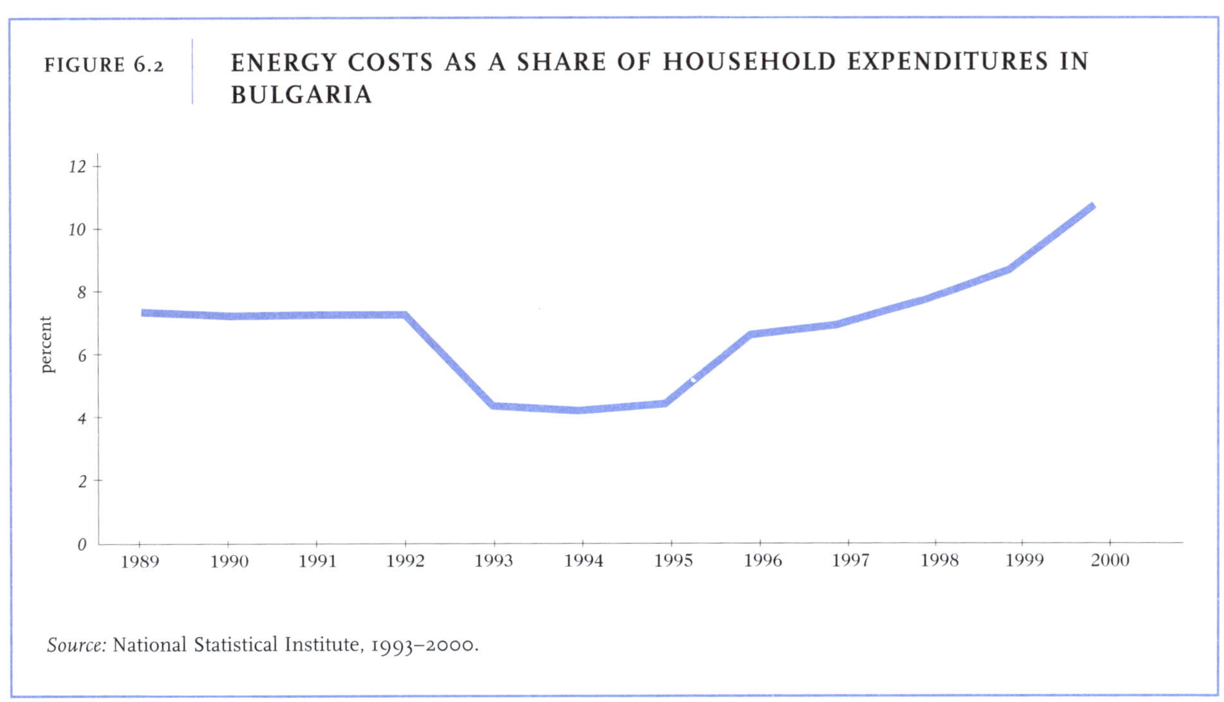

Source: National Statistical Institute, 1993–2000.

for Central and Eastern Europe and the World Resources Institute, 2001).

Late in the reform process, the World Bank emerged as a strong advocate for increased efficiency, and has demonstrated that efficient use is an alternative to rampant expansion of supply potential. However, this effort has come late in the process, and until recently this message has had little domestic support. A focus on energy efficiency has been a politically inconvenient message, since it threatens to undermine the interests of those who advocate continued capacity expansion. A domestic political constituency for energy efficiency is necessary to counter the influence of forces supporting enhanced supply. Fortunately, early signs from the new government, as suggested by the emphasis on energy efficiency in the 2002 Energy Strategy, suggests that such a constituency may in fact be forming.

The approach to social issues has followed a similar trajectory. Despite indications of a serious social problem, no assessments or forecast of the social burden of reforms were initiated in the early stages

of reform. The early evidence suggests that reforms in the electricity sector will substantially increase both electricity and district heating prices for households. Since the initiation of reforms, household budgets for energy have steadily increased, rising from a 7 percent share in 1996 to 11 percent in 2000. *(See Figure 6.2.)* The burden is heaviest for the poorest households. For example, old-age pensioners paid an average of 14 percent of their household budget on energy costs, the second largest category of household expenditure (Dimov, 2000).

For several years, the government's approach has been to postpone seeking a solution, or to seek short-term fixes. In order to put the sector on a sound financial footing, the IMF has relentlessly urged price increases, but without providing any solution to the social cost of price these increases. The government negotiated a freeze on electricity and district heating prices with the IMF for two years beginning in 2000. Moreover, it has instituted a program for targeted energy support during the cold season, which was extended to a substantial 19 percent of households (12 percent of the population) (Dimov

2000). This is clearly a long-term problem not amenable to a short-term fix, as indicated by the fact that energy prices in Bulgaria are rising toward Europe-wide prices as the country increasingly seeks to integrate its markets, yet Bulgarian incomes are around 30 percent of European incomes. Hence, for a large segment of the population, energy costs are likely to be a considerable burden for some time to come.

In the Bulgarian context, the link between social and environmental outcomes is particularly apparent. End-use efficiency and household conservation efforts could bring substantial social benefits by buffering the cost of energy price increases, even while providing environmental benefits. Hence, greater efficiency provides a positive link between environmental and social outcomes. The potential negative linkages are also clear. For example, with rising district heating prices, many consumers have disconnected from the system, calling its financial viability into question. Instead, customers have switched to electricity for heating since it allows them to control their heat use and associated expenses. The environmental cost of this switch is significant, since district heating is based on natural gas combustion, while a considerable share of electricity is generated from coal.

More recently, the World Bank's preparation of an adjustment loan seeks to build on its studies demonstrating the considerable benefits of energy efficiency on both social and environmental grounds. In addition, the World Bank has revived earlier projects to rehabilitate the district heating system in major cities. Hence, there are signs that environmental and social factors may yet receive explicit attention.

In sum, despite the considerable environmental and social stakes in the reform process, successive governments throughout the 1990s made little effort to assess the implications of electricity sector reforms on these broader concerns. Among donor agencies, the IMF has maintained its support for the rapid introduction of market prices. The World Bank advocated attention to energy efficiency and social protection, but only after the government's initial round of reform design and implementation. The slow pace of reform, combined with a new government, may yet provide an opportunity to correct the earlier inattention to public benefits in the reform process.

CONCLUSION

It took a crisis situation for reforms in the electric power sector to be undertaken in Bulgaria. Efforts at reform funded by a range of donor agencies in the early 1990s resulted in a series of technical studies, but few actual steps in this direction. Only pressures from the IMF led the Bulgarian government to take serious steps toward reform. However, in a crisis context, the Bulgarian government had limited ability to frame the approach to reform, which was largely dictated by the IMF's concern with financial transparency, introduction of commercial principles in the sector, tariff increases, and ultimately privatization.

Once in the implementation stage, domestic actors and particularly the SAEER (the former Committee of Energy) reasserted their authority, but not necessarily in the public interest. Reform implementation led to a flawed sequencing of reform, where unbundling was given priority over establishment of an institutional framework. Moreover, the government's supply orientation and preoccupation with developing an export market for electricity continued unabated. Despite donor leverage in the initial stages, the IMF in particular was unable to change the government's course in these two important areas.

Neither the government nor donor agencies brought sufficient attention to bear on concerns of public benefits in the design stage of the reforms. In the midst of a financial crisis, the IMF sought attention to the financial dimensions of reform. While well aware of the social costs of tariff increases, the IMF made no attempt to examine ways of insulating the population in the long run from the social costs of price increases. While the World Bank had conducted studies of environmental issues in the sector in the early 1990s, there is little evidence that these studies shaped the design of reforms carried out in the late 1990s. Through subsequent studies, the World Bank did criticize the government's supply-oriented mindset obliquely through the avenue of environmental concern, by demonstrating the potential of greater efficiency to obviate the need for supply increases. By the time this and other important new studies on the environmental and social dimensions of reform were complete, however, political momentum had increased in favor of supply expansion. However, the election of a new government and the preparation of a World Bank loan for structural reform provide scope to change course, and World Bank studies on the energy-environment link may yet prove to be a useful tool.

The experience in Bulgaria also calls into question the effectiveness of coordination among the donor agencies. While the World Bank was active in the sector in the early 1990s, there is little evidence that this knowledge played a role in shaping IMF-led reforms a few years later. Moreover, two years into reforms, there was a growing wedge between the positions of the two institutions, as the IMF accepted the government's interest in developing new capacity through joint ventures, while the World Bank continued to find fault with this approach. Only in 2001 did the two institutions better coordinate their message to the government.

Among public benefit concerns, social issues are at the top of the political agenda. Price increases in electricity and in district heating are the greatest public concern, and also occupy the political attention of the government. While environmental issues are of significant international concern, particularly

given environmental standards necessary for EU accession, they have not risen to an equivalent level of political importance.

Neither social nor environmental concerns have been articulated in open debate over reforms. Indeed, reforms have been dominated by a small group of government bureaucrats and donor agencies, with little scope for public debate. The lack of open debate through the 1990s may have contributed to glaring disconnects in the reform process, such as that between the goal of increasing commercial discipline and the government's insistence on new capacity addition for exports, which appeared doomed to be loss-making. With the reform process having entered a new phase in 2001, there is still time for a supply-driven approach to reforms to give way to one motivated by concern over the broader public interest, and for more attention to be paid to the considerable potential for social and environmental gains from energy efficiency in Bulgaria's electricity sector.

NOTES

1. Dimitar Doukov wishes to thank colleagues at EnEffect, in particular, Mr. Atanas Stoykov who provided information and advice at all stages of the project. He also wishes to acknowledge the advice and contribution of Mr. Lulin Radulov. This chapter draws on Dimitar Doukov, 2001, "Bulgaria Power Sector Reform." Unpublished paper, Eneffect, Sofia, Bulgaira (July). Online at http://www.wri.org/governance/iffepowercases.html.

2. The effects of this stabilization program can be traced through data produced by the National Statistical Institute (1993-2000).

3. Calculated by Doukov from National Statistical Institute (1993-1997).

4. See, for example, European Parliament (1997), which states that "Integrated electricity undertakings shall, in

their internal accounting, keep separate accountings for their generation, transmission and distribution activities."

5. Interview with former government bureaucrat, August 2000. All interviews for this chapter were conducted on a not-for-attribution basis. Consequently, interviewees are identified only by their institutional affiliation.

6. See European Union Electricity Directives: Resolution 98/C4/01 (18 December 1997); COM(97) 599; Decision 99/21/EC, Decision 96/737/EC; Decision 2000/646/EC.

7. Republic of Bulgaria Council of Ministers (1998a) envisions decommissioning of Units 1 and 2 of Kozloduy nuclear power project in 2004 and 2005. According to the Memorandum signed on November 29, 1999 between Bulgaria and the European Commission it has been agreed that the decommissioning of Nuclear Units 1 and 2 of Kozloduy NPP should be carried out by 2003.

8. For example, in 1999, targeted assistance was extended to 19% of households, or 12% of the population, indicating the large segment of the population that requires assistance for basic heating.

9. Interview with World Bank official, April 14, 2000.

10. This was clearly stated in a Memorandum between the Government of Bulgaria and the IMF, February 2000. Also see *Capital Newpaper* (2000).

11. It is relevant here to note that the Chairman of the Energy Committee in Parliament was not from the ruling party.

12. Interview with World Bank staff member, April 14, 2000.

13. Personal communication, World Bank staff, April 2002.

14. Personal communication, World Bank staff, April 2002.

REFERENCES

Alexandrova, Galina. 1997. "Mr. Shilyashki Messed up the Energy Bills." *Capital Newspaper*. Issue 18. (May).

———. 2000. : "The offshore company "3?" and NEC bargained again unprofitable contract." *Capital Newspaper*. Issue 28 (July).

Capital. 2000. "IMF: We Postpone the Division of NEK." *The Capital*. (February 7).

Center for the Study of Democracy. 1995. *Energy Policy in the Context of Bulgaria's Accession to the European Union*. Sofia: Center for the Study of Democracy.

European Parliament. Article 14, item 3., Directive 96/92/ EC of the European Parliament and of the Council of 19 December 1996. Official Journal of the European Communities No L 17/20, 30. 1. 97.

Dimov, Martin. 2000. "Social Aspects of the Energy Sector Reform." Unpublished manuscript. Study contracted by EnEffect. Sofia: EnEffect.

Energoproekt and Sofia Energy Centre. 2001. *Determination of Market Potential for Small Boilers on Fluidized Bed Combustion, Pre-feasibility Study*. Sofia: Energoproekt and Sofia Energy Centre.

Energy Charter Secretariat. 1996. *Energy Charter Treaty and Related Documents*. Brussels: Synergy Programme of the European Union.

Kahkonen, Juha. 2000. Letter from IMF representative Juha Kahkonen to Deputy Prime Minister and Minster of Economic Affairs Peter Zhotev, April 27, 2000.

National Electric Company. 2001. *Annual Report 2000*. Sofia: National Electric Company.

National Statistical Institute of the Republic of Bulgaria. 1993-2000. *Statistical Yearbook of the Republic of Bulgaria*. Sofia: National Statistical Institute.

———. 2002. *Statistical Reference Book of the Republic of Bulgaria 2001*. Sofia: National Statistical Institute.

Pari Daily. 2000. "The Ideas of Shilyashki Will Overthrow the Government." *Pari Daily* (April 14):73.

Regional Environmental Center for Central and Eastern Europe and the World Resources Institute. 2001. *Good Practices in Policies and Measures for Climate Change Mitigation: A Central and Eastern European Perspective*. Szentendre: Regional Environmental Center for Central and Eastern Europe and the World Resources Institute.

Republic of Bulgaria Council of Ministers. 2000a. *National Climate Change Action Plan*. Sofia: Council of Ministers.

———. 2000b. *Memorandum on Economic Policies of the Government of the Republic of Bulgaria*. Online at: http:// www.imf.org (March 9, 2002).

———. 1998a. *National Strategy for Development of Energy and Energy Efficiency til 2010*. Sofia: Republic of Bulgaria.

———. 1998b. *The Strategy for Development of the Energy Sector by 2010*. Sofia: Republic of Bulgaria.

Republic of Bulgaria National Assembly. 1999. "Law on Energy and Energy Efficiency." *State Gazette*. (July 16): 64.

———. 1998. *"Law of State Budget of the Republic of Bulgaria for 1999." State Gazette* (December 29):155. Paragraph 9, Annex 4.

Todorov, Vasil. 2001. "The Economy: Will We Succeed? The Diagnosis is Before Therapy." *Pari Daily*. Issue 144 (August 1): 7.

World Bank. 1992a. "Bulgaria Environment Strategy Study, Report No. 10142-BUL." Washington, D.C.: World Bank.

———. 1992b. "Bulgaria Energy Strategy Study, Report No. 10143-BUL." Washington, D.C.: World Bank.

———. 1993a. "Bulgaria Energy Project, Report No. 11250-BUL." Washington, D.C.: World Bank.

———. 1993b. "Bulgaria Energy Project." Report No. 11250-BUL." Washington, D.C.: World Bank.

———. 1994. "Bulgaria Environment Strategy: Update and Next Steps, Report No. 13493-BUL." Washington, D.C.: World Bank.

7

GHANA

ACHIEVING PUBLIC BENEFITS BY DEFAULT

Ishmael Edjekumhene
Navroz K. Dubash[1]

INTRODUCTION

As with many other countries in the developing world, the slow shift from public to private control in Ghana's electricity sector is part of a wider move toward a market economy in the aftermath of a decade of economic turmoil. In contrast to many of the other cases in this study, Ghana has a relatively small electricity sector, consisting mainly of two large public corporations. Reform in Ghana has posed both institutional and technical challenges. Of particular concern is expanding the population's access to electricity. The World Bank has historically played a dominant role in the Ghanaian electricity sector, and has been instrumental in the reform process. But the government of Ghana (GOG) has been firm in seizing ownership of its reform program and independently directing the course of reforms. For a profile of the electricity sector in Ghana see Box 7.1.

BACKGROUND

State-Owned Enterprises Control the Commanding Heights

From the date of its independence in 1957 until the late 1980s, Ghana pursued a state-led economic development strategy. Central to this vision was the creation of strong state enterprises to develop infrastructure and public services. In the post-colonial vision of development, which Ghana shared at the time, this strategy of publicly provided services formed the backbone of an industrializing nation.

<table>
<tr><td>BOX 7.1</td><td>PROFILE OF THE ELECTRICITY SECTOR IN GHANA</td></tr>
</table>

Population (2001)[1] : 18.5 million

Population with access to electricity (2000) [2]:
Total: 40%   *Rural: 17%*   *Urban: 77%*

Installed electricity generation capacity (2001) [3]
  Total: 1.6 gigawatts (0.05% of total world capacity)
  Thermal: 35%
  Hydro: 65%
  Nuclear: 0
  Geothermal and Other: 0

CO_2 emissions from electricity and heat as a share of national emissions (1999)[4]: 23%

Notes:

1. Ghana Statistical Service, 2002. "2000 Population Census Report". Accra.

2. Ministry of Energy, 2001. "Energy for Poverty Alleviation and Economic Growth: Policy Framework, Programmes and Projects". November, 2001. Accra.

3. www.eia.doe.gov/pub/international/eiapdf/to6_04.pdf (February 6, 2002).

4. Computed by WRI using International Energy Agency (IEA) data. IEA, 2001. *CO2 Emissions from Fossil Fuel Combustion.* Paris: OECD.

By the early 1980s, state-owned enterprises in
Ghana had turned in a poor financial performance
for several consecutive years. The resultant losses
were financed through borrowing or taxes, which
imposed a debt burden and a misallocation of
resources (Edjekumhene, 2000). In 1982, state-
owned enterprises racked up operating deficits
amounting to over 3 percent of GDP, an amount
almost as large as total spending on education,
health, social security, and welfare (Berenschot and
Bosboom, 1995).

The problem was worsened by external shocks. A
downturn in the world price of cocoa—Ghana's
major export—combined with lax fiscal and monetary policies led to a prolonged period of crisis
accompanied by severe inflation. Between 1980 and
1983, GDP declined by 17 percent and export earnings by over 53 percent, while external debt rose by 17
percent (Kapur et al., 1991).

This crisis led the government to negotiate an
"Economic Recovery Program" with the Bretton
Woods institutions in 1983, which subsequently
evolved into a more complete structural adjustment
program in 1986. The adjustment program implemented state retrenchment, a floating exchange rate,
price liberalization, and privatization of state-owned
enterprises (Partiff, 1995). As a large and economically significant sector, the electricity sector was high
on the list for attention (Opam and Turkson, 2000).

THE PATH TOWARD REFORM: THE PRE-REFORM STRUCTURE AND DIAGNOSIS OF PROBLEMS

In keeping with the conventional wisdom prior to the
1990s, Ghana's electricity sector has long been
organized around public utilities. The Volta River
Authority (VRA) generates almost all of Ghana's
electricity through two large hydroelectric projects
and two recently installed thermal power plants. VRA
is also responsible for the transmission grid. The
Electricity Company of Ghana (ECG) is the primary
distribution utility. The sector is governed by the
Ministry of Energy (formerly the Ministry of Mines
and Energy, or MOME), which has responsibility for
policymaking and coordination. In addition, a State
Enterprises Commission was established in 1987 to
develop performance contracts for state-owned
enterprises, including VRA and ECG.

However, these institutions have failed to ensure
widespread access to electricity. By 1993, only 24
percent of the population was served by electricity
(World Bank, 1993a). In 1989, the government
instituted a National Electrification Scheme aimed at
expanding access to the entire population by 2020.
This ambitious goal remains a critical part of government policy in the sector.

Throughout its history, the electricity sector in
Ghana has received considerable financial support
from international donors, led by the World Bank.
The World Bank provided almost half the external
loan component for the country's first hydroelectric
dam and substantial portions of subsequent projects.
Between 1961 and 1995, the World Bank conducted
eight lending operations aimed at the power sector.
The institution has been the single most important
financier and a critical catalyst of additional international public support for technology development,
institutional development, and management reform.
In particular, the World Bank has been central in
building VRA and ECG—the two institutional
building blocks of the sector. Understanding the
genesis of problems in the sector requires delving
into the past record of each of these institutions. An
overview of the pre-reform sector is provided in
Figure 7.1.

Volta River Authority: The Favored Child

With the personal and strong support of then-
President Nkrumah, VRA was established in 1961. It

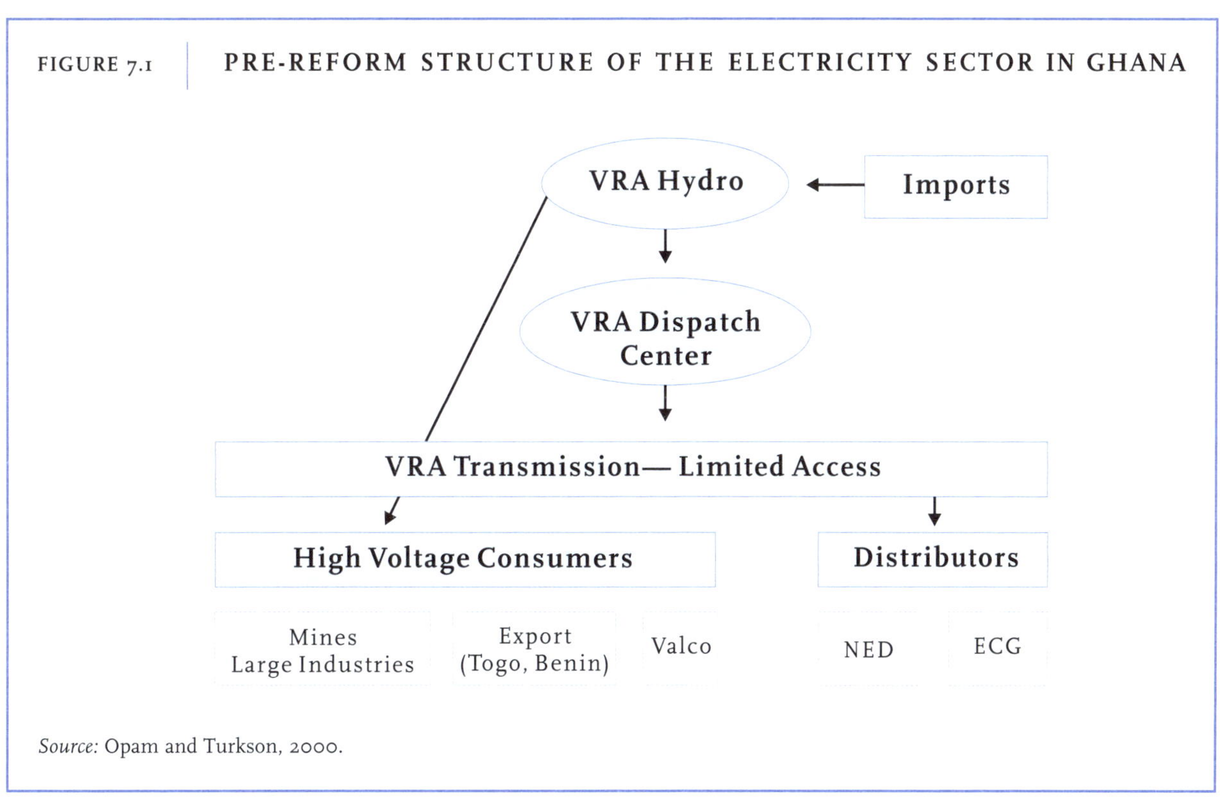

Source: Opam and Turkson, 2000.

was inspired by the example of the Tennessee Valley Authority in the United States. VRA, with a total capacity of 1,072 megawatts through most of the 1980s and 90s, has operated as a public monopoly in generation based on two large hydroelectric projects. It controls the bulk power sales market and has a monopoly in transmission services (World Bank, 1993a). It has also been the monopoly supplier of electricity to the distribution company, the Electricity Corporation of Ghana. The central role of VRA, and the conditions of its creation, has shaped recent reform efforts in several significant ways.

The government created VRA as an independent entity with separate governance and legal structures. This caused both management complications and political resentments (Brew-Hammond, 1994). These arrangements, critics have argued, crippled the force of the Electricity Act—passed around the same time as the creation of VRA—and the authority of what was then the Electricity Department under the Ministry of Public Works, later the Ministry of Mines and Energy.[2]

VRA was run for a long time as an economic enclave, insulated from problems that have beset other parts of the economy. When VRA was established in 1961 with an initial capacity of 588 megawatts (Brew-Hammond, 1994), total demand in Ghana was barely 100 megawatts. To absorb the surplus electricity, VRA agreed to supply electricity to industrial consumers, prominent among them the privately owned Volta Aluminum Company (VALCO). As recently as 1991, VALCO consumed 59 percent of Ghana's total power supply (Opam and Turkson, 2000).[3] In later years, VRA's customer base expanded to include additional industrial consumers such as mines, and exports of electricity to neighboring countries such as Côte d'Ivoire, Togo, and Benin.[4] As a result, VRA received a large share— estimated in 1990 at 70 percent—of its revenues in foreign exchange. This insulated VRA from the effects of the periodic and dramatic currency devalu-ations of the 1980s and 1990s (World Bank, 1990).

VRA has won plaudits for its strong financial track record from external actors such as the World Bank, which described it as "a relatively well-run public utility with few institutional and financial problems" (World Bank, 1995, p. 2). Yet, critics contend that its privileged status as an enclave, which allowed it to sell predominantly to wealthy industries that pay in foreign exchange, is a large part of the explanation for its relative success.[5] Indeed, with its emphasis on commercial profitability, VRA had little incentive to extend its transmission network to serve the Ghana-ian people, who are both less lucrative and more difficult customers to serve than giant industrial consumers. If VRA has been successful, contend the critics, it is because success has been defined in terms of commercial viability rather than contribu-tion to development goals.

Backed by support from external donors, VRA's role continued to grow. In 1987, its mandate was amended to include distribution of electricity in the northern zone of the country through establishment of a Northern Electricity Department (NED) within VRA. Moreover, VRA was envisioned as playing a central role in promoting opportunities for regional cooperation in power development and exchange in West Africa (World Bank, 1990).

Electricity Company of Ghana: The Step-Child

The Electricity Company of Ghana (ECG), established as the Electricity Corporation of Ghana by a govern-ment decree in 1967, was intended to be the main distribution entity in Ghana. ECG was charged with the bulk purchase of electricity from VRA for

distribution to all categories of consumers, with the exception of VALCO and other large industrial consumers supplied directly by VRA.

The establishment decree required ECG to conduct its operations on a commercial basis. However, ECG has lurched from crisis to crisis through the 1980s and early 1990s, defying several efforts to set it on the right footing. Following the broader economic crisis of the 1980s, the situation was worsened. As losses mounted, ECG developed a negative total equity, a high debt-to-equity ratio, and a heavy debt-service burden (World Bank, 1993a).

Several efforts to improve this situation—including a twinning arrangement with an Irish utility, a World Bank-supported project to restructure financially, and the discipline of performance contracts—have not met with success. A central barrier to ECG's viability is a tariff structure that is too low to ensure its financial health (Opam and Turkson, 2000). In particular, tariffs have not kept pace with currency depreciation, which raises ECG's costs in loan interest and exchange rate fluctuation charges. Attempts to change and revise upward the tariff structure have been unable to deal with the funda-mental problem—the Government of Ghana was hard pressed to politically sustain tariff hikes when electricity service was both unreliable and of low quality.

Most recently, the World Bank included a substan-tial component of institutional strengthening in its National Electrification Project. In the words of the project document, "VRA has been a relatively autonomous, financially viable, and well-run entity; attention is now focused upon bringing ECG to a similar standard" (World Bank, 1993a, p. 5). The loan agreement required the government to agree to a self-adjusting tariff formula to guarantee a minimum rate of return on assets. Even more significant, ECG was required to create a separate Directorate of Customer Services for ECG, to be operated by Electricité de France under a performance-based management contract (World Bank, 1993a, p. 7). This approach partially mirrored one recently adopted in neighboring Côte d'Ivoire, and was also an indication of an emerging World Bank policy formalized in 1993 to encourage importation of services in the energy sector (World Bank, 1993b). (*See Box 7.2.*)

Even as these efforts were put in place, ECG was already slated for conversion to a public liability company as part of a commitment made to the World Bank under the 1980s structural adjustment program. Notably, VRA was not included on this list, since it was not perceived as an underperforming enterprise (Opam and Turkson, 2000). By the mid-1990s, the future of ECG was already in doubt.

The Seeds of Reform

This thumbnail sketch of Ghana's electricity sector suggests several issues that have both forced and shaped reforms in the sector. The role of VRA is a source of contention. To some, VRA is the saving grace of the sector. To others, its enclave nature, which allows it to maintain financial viability, is an obstacle to a sensibly integrated sector. ECG's struggles to maintain financial viability, and the difficulty of doing so when confronted by declining real tariffs (adjusted for depreciation), have left it in a bind. By the early 1990s, the World Bank's growing preference for the importation of services and an expansion of private sector engagement threatened to undermine support for ECG. And the situation was further complicated by broader efforts by the Government of Ghana to restructure the economy in keeping with the structural adjustment policies of the 1980s. Collectively, these features of the sector were the seeds of reform.

The need for change was brought into focus by a set of three circumstances that developed through the 1980s, two of which were internal and one external to Ghana. First, since the country relied on hydropower for 95 percent of its capacity, a severe drought in 1982-83 (and again in 1993) led to a severe electricity shortfall. By 1984, generation had reached 36 percent of its pre-drought 1982 levels, and only recovered to pre-drought levels in 1989 (Opam and Turkson, 2000). Adjustment to these lower levels of generation was compounded by VRA contracts to supply its industrial users, notably VALCO.[6]

Second, between 1985 and 1993, domestic demand for electricity rose at a substantial 10.8 percent per year, increasing to 15 percent between 1993 and 1995 (Ministry of Mines and Energy, 1996). Growth in demand was caused by an increase in economic growth following the economic crisis of the early 1980s, as well as the requirements of the National Electrification Scheme. By 1994, the flow of electricity sales to Côte d'Ivoire had been reversed, as Ghana became a buyer rather than a seller of electricity. The country was thus caught between growing demand and shrinking supply.

To address the shortfall, the government needed to urgently raise funds for new generation capacity to meet contractual obligations to VALCO, export demands to neighboring Togo and Benin, and to service the current and expanding projected needs of its own consumers. The government also needed funds to improve existing transmission and distribution infrastructure.

Third, and perhaps most important, the traditional source of financing for the power sector was drying up. In 1993, a new World Bank policy of "commitment lending" required sectoral reform as a precondition to further loans. The Bank said that it would not finance VRA's expansion program for a thermal power project at Takoradi without basic structural reforms.

In sum, the growing electricity shortfall ensured that the broader structural problems in the sector could no longer be ignored. However, the World Bank policy of "commitment lending" was the proximate cause that forced action. This view was widely held among GOG officials involved in designing the reforms. As one senior official engaged in the reform succinctly put it, "if the World Bank had given us money, no one would have talked about reform."[7]

CONSTRUCTING A REFORMED ELECTRICITY SECTOR

That various actors relevant to Ghana's electricity sector had come together on the need for reforms did not mean that they all agreed on the direction of those reforms. They actually differed considerably on the scope of the reforms and the vision of a post-reform sector. In this section, we examine how the views espoused in the debate by the World Bank, VRA, and the government combined to shape reforms. The chronology of the reform process is laid out in Box 7.3.

Contrasting Views on Reform

The World Bank's thinking during this period is encapsulated in a review of lending for electricity in sub-Saharan Africa that was written by its Operations Evaluation Department after the 1993 World Bank policy was put in place (Covarrubias, 1996). The review argues against the unbundling of the sector—as was carried out in Chile and the United Kingdom—due to the small size of the power sector in

short-term political interest—to avoid electricity shortages in the run-up to elections. Significantly, in a major concession to investors, the government decided to provide fuel to new IPPs at no cost.

No regulatory reforms preceded privatization, but there was a subsequent flurry of regulatory activity. A number of new institutions were set up, including bodies to supervise EECI, address financial issues, address engineering issues, develop a national policy for electricity, and promote rural electrification. There was a simultaneous reshuffling of responsibilities, often with overlapping jurisdictions. As a result, regulation, planning, and policymaking within the sector became increasingly duplicative and unclear. As one observer noted, "each private operator can literally pick the government body with which it is comfortable in order to solve its problem with the lowest possible risk."

As a result of the existing confusion, a new round of reforms is in the cards that will aim to reorganize the system from a vertically integrated monopoly to an unbundled system. However, the government is locked into the existing agreement with CIE, which expires in 2005 and poses problems for any future reform.

Sources:

Brew-Hammond, Abeeku. 1997. "Ghana and its CCGT." *Financial Times Energy Economist.* September 197 191/13.

Berg, Elliot, Patrick Guillaumont, Jacky Amprou, and Jacques Pegatienan. 1999. "Côte D'Ivoire." In *Aid and Reform in Africa.* Edited by Shantayanan Devarajan, David R. Dollar, and Torgny Holmgren. Washington, D.C.: World Bank. Online at: http://www.worldbank.org/research/aid/africa/cotedivoire.pdf (May 30, 2002).

N'Guessan, Etienne K. 2000. "Privatization of the Power Sector in Côte d'Ivoire." In *Power Sector Reform in Africa.* Edited by John K. Turkson. London: MacMillan Press.

World Bank. *African Gas Initiative: Main Report (Volume 1).* Energy Sector Management Assistance Programme. ESM 240. February 2001.

most sub-Saharan countries and the inadequate and weak regulatory frameworks in place. Instead, the report recommends maintaining a dominant generator, with selective reliance on independent power producers (IPPs) as an interim measure until conditions are ripe for competitive markets. The single most forceful recommendation is to focus on distribution reform as the source of problems related to cost recovery and service delivery. Here, pointing to the Côte d'Ivoire example (*see Box 7.2*), the report suggests the use of management contracts—potentially extending to outright distribution concessions—which would make concessionaires responsible for ongoing investments. It also urges the introduction of transparent regulation, noting that nonenforceable performance contracts have proved largely unsuccessful; suggests consideration of decentralized community-based distribution; and

favors reform of purchase tariffs. Finally, the review notes the political challenges in shifting away from a public monopoly model and recognizes the central importance of "borrower ownership" of a reform process.[8]

The country was caught between growing demand and shrinking supply.

These ideas were put to the test in negotiations between the World Bank's International Development Association (IDA) and MOME over a credit for the Takoradi thermal power plant. While not specifying a particular reform model, the World Bank conveyed a desire to maintain VRA as an intact

CHRONOLOGY OF ELECTRICITY SECTOR REFORM IN GHANA

January 1994	GOG issued a Strategic Framework for Power Sector Development Policy.
March 1994	Ministry of Mines and Energy (MOME) engages a consultant (SYNEX of Santiago, Chile) to study the opportunities for restructuring the power sector.
June 1994	SYNEX submits report, which proposes a new power market for the country.
June 1994	Preparation of a sector policy letter by the MOME, which laid out sector objectives, institutional guidelines, and regulatory principles.
September 1994	Establishment of the Power Sector Reform Committee (PSRC) by the MOME to coordinate the design and implementation of reforms.
January 1995	World Bank approval of a loan for a thermal power project at Takoradi, with a condition for establishment of a committee on power sector reform.
Mid-1995	Formation of two Task Forces by PSRC: Task Force I on operational technicalities of the reform program, particularly pricing and commercial organization of the power market; Task Force II on the legal implications of the proposals for the reform.
August 1996	Stakeholders' workshop to discuss proposals from the Task Forces.
September 1996	Formation of Review Task Forces on distribution and customer service to address specific issues emerging from the stakeholders' workshop.
February 1997	Electricity Company of Ghana registered as a limited liability company to take over the assets and operation of the Electricity Corporation of Ghana.
April 1997	PSRC submits a summary report to the government.
May 1997	Establishment of Power Sector Reform Implementation Secretariat to coordinate implementation of the recommendations contained in the report.
October 1997	Enactment of the Public Utilities Regulatory Commission Act, 1997 (Act 538), which establishes Public Utilities Regulatory Commission (PURC).
December 1997	Enactment of Energy Commission Act, 1997 (Act 541), which established the Energy Commission.
September 1998	PURC approves tariff increase for all categories of consumers.
July 2000	MOME submits draft Electricity Regulation to Parliament for approval.

generation entity, to supplement generation capacity with IPPs, and focus on reform of ECG.[9] In the subsequent phrasing of the loan document, the World Bank stated that "the most appropriate role for the private sector at this time is through participation in a performance-based management contract for the proposed thermal power project" (World Bank, 1995, p. 7). A condition for the loan was that the GOG establish a committee on power sector reform with agreed-upon terms of reference, and that this committee issue recommendations for the sector within a specified time. Also included in the loan agreement for the Takoradi power plant was a component for support of VRA's institutional development, suggesting that VRA would remain an intact public entity for some time to come.

Perhaps surprisingly, the GOG proposed a considerably more far-reaching set of reforms. Having agreed with the World Bank, in principle, on the need to reform the sector, the Minister of Mines and Energy, and the Minister of Finance prepared a letter to the World Bank in 1994 laying out a "Strategic Framework for Ghana Power Sector Development Policy." This document signaled that "the Government will be introducing some fundamental reforms to establish the conditions in the electricity sector for greater operation efficiency and competition, private sector participation, and the development of an arm's length approach to Government regulation of power sector entities" (Peprah and Botchwey, 1994). As the summary in Box 7.4 suggests, the document promised far-reaching change in the sector, including the introduction of competition and a focus on limiting sovereign guarantees.

Moreover, the GOG took the unusual step of hiring its own consultant independently from the World Bank—SYNEX Consulting Engineers from Chile—to flesh out its reform ideas. As the World Bank report states, "MOME took the initiative of developing a comprehensive new policy framework" (World Bank, 1995). Specifically, the GOG charged SYNEX with evaluating the key issues and examining the options for applying a market-oriented approach in the sector. The choice of a consulting firm from Chile was significant, since Chile was among the first countries to introduce a market approach in electricity. A Chilean firm could be expected to advocate a

---

BOX 7.4 | **SUMMARY OF STRATEGIC FRAMEWORK FOR POWER SECTOR DEVELOPMENT POLICY**

Long-term strategic goals

- assure reliable, economically efficient, and equitable supply of electricity to meet the country's growing needs for socio-economic development;

- serve as a basis for creating attractive, marketable assets; and

- develop an efficient mix of commercially viable public and private sector utilities.

Policy objectives

- increase management accountability in the existing public utilities;

- move the power sector away from the existing monopolistic structure toward a more decentralized structure that would expose the public utilities to competition in both generation and distribution;

- encourage private sector investment through the establishment of IPP schemes and the provision of an "open access" grid to facilitate direct sales by IPPs to consumers;

- reduce the extent to which the government is called upon to provide sovereign guarantees; and

- establish a transparent regulatory framework to enable competition in the sector.

Sources:

Ministry of Mines and Energy. 1994. "Strategic Framework for Power Sector Development Policy." Accra, Ghana

Peprah, Kwame, and Kwesi Botchwey. 1994. "Strategic Framework for Power Sector Development Policy." Letter to Joanne Salop. World Bank. (March 23).

similar approach for Ghana—an approach the World Bank review had explicitly rejected as premature for sub-Saharan Africa.

Based on this study, the GOG established a Power Sector Reform Committee (PSRC) to work out the design and implementation of reforms. The PSRC consisted of eight members drawn from MOME, VRA, ECG, and the private sector. It organized its work around two task forces, on the pricing and commercial organization of the electricity market, and the legal implications of the reform proposals. The results of the task forces were reviewed at a workshop in August 1996. Participants at the workshop included business representatives from Ghana and abroad; large industrial consumers such as VALCO, the World Bank, and other donor agencies; and a range of resource persons, including representatives of utilities from the United States and Côte d'Ivoire. Three additional task forces on a grid code, electricity distribution, and customer service were added to respond to issues raised during the workshop (Opam and Turkson, 2000).

The Architecture of the Post-Reform Sector[10]

The PSRC submitted a final report to the GOG in April 1997, with substantial recommendations to transform the electricity sector in Ghana. These recommendations were adopted by the Cabinet, which created an implementation secretariat to proceed with reforming the sector. The vision of the transformed sector is described below under the general categories of generation, transmission, distribution, market coordination, and regulatory reforms and shown in Figure 7.2.

Generation

Electricity generation was opened up to generators other than VRA. It was not recommended that VRA itself be privatized, but that VRA, as with other utilities, be re-organized into "strategic business units" as part of a broader effort to improve management accountability. The core business of hydro generation was left untouched. The PSRC also recommended recapitalization through public-private partnerships. Generators can sell power directly to large consumers in wholesale markets. Hence, each generator had three options: to trade power to other generators, to large consumers directly, or to distribution companies, who would then serve final consumers.

Transmission

The transmission system was to be unbundled from VRA and established as a separate publicly owned national grid company on an open access or nondis-criminatory basis. Large consumers were to be given access to the transmission network to participate in the wholesale market.

Distribution

The PSRC recommended the establishment of both a deregulated wholesale market for consumers with demand larger than 5 megawatts and a regulated market for smaller consumers. Large consumers were to contract directly with generators. For con-sumers smaller than 5 megawatts, the country was divided into five distribution zones. Four of these were based on areas formerly served by ECG, and one was based on the regions served by NED. ECG was subsequently transformed into a limited liability company with autonomous distribution concessions. NED was to be unbundled from VRA to serve as a separate distribution concession. These concessions would eventually be privatized.

Market Coordination

A new Ghana Economic Load Dispatch Centre (GELDIC) would be given authority over coordination and dispatch in the system. Essentially, GELDIC would be responsible for ensuring that physical and financial transactions were coordinated in a manner that ensured the system's reliability, in keeping with the principle of minimum operation cost.

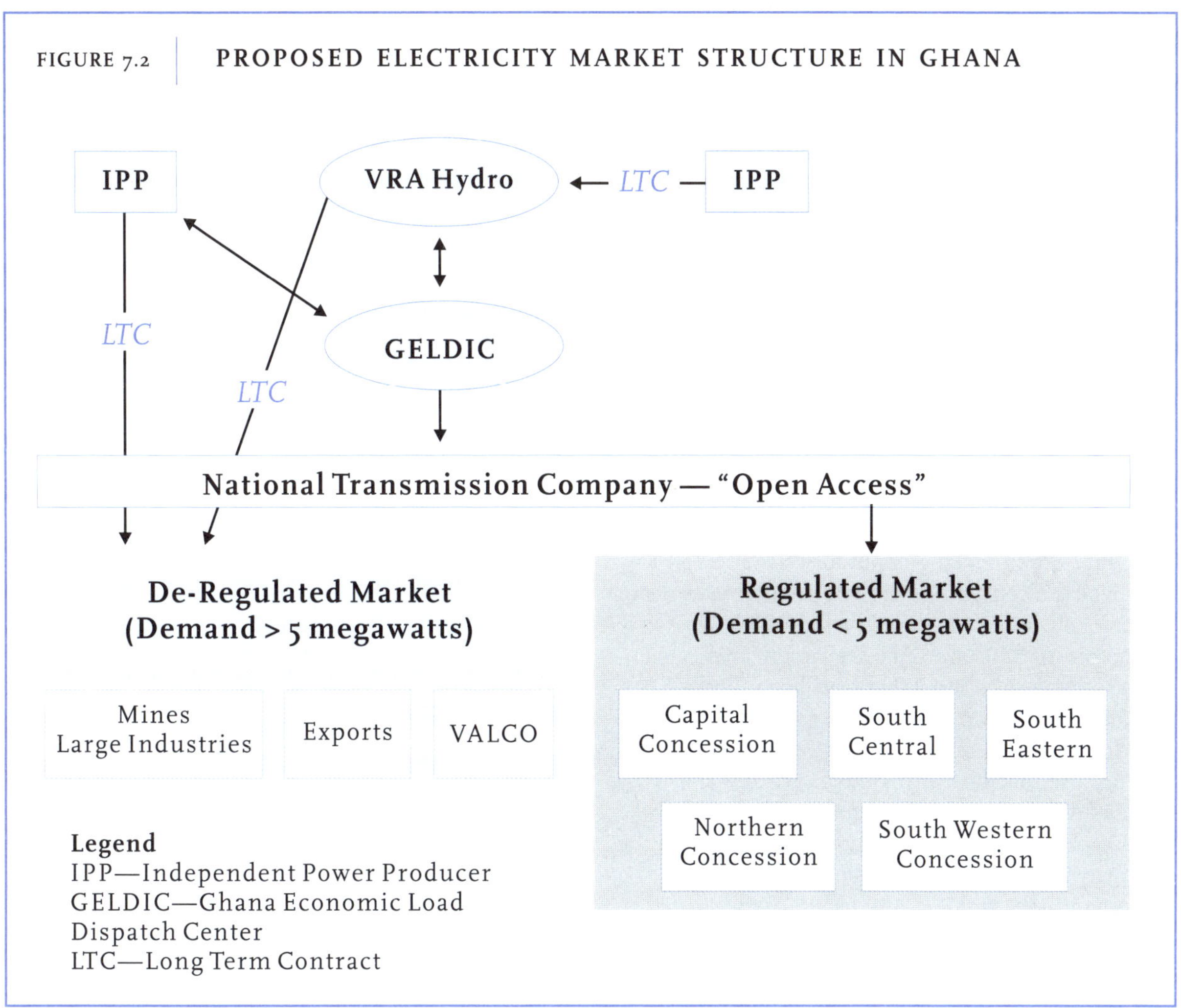

Regulatory Framework

Perhaps most crucially, the PSRC recommended far-reaching regulatory reforms. The cabinet accepted the PSRC recommendation for an independent regulatory body, but implemented this recommendation through a two-tier framework. First, in keeping with the PSRC, they established a Public Utilities Regulatory Commission (PURC) in 1997. Second, they also established an Energy Commission, whose functions include providing advisory services to the MOME. While independent of the Ministry, the commission is also subject to direction by the minister as necessary to ensure the public interest—

a provision based on a constitutional requirement that key sectors should be supervised by a technical commission.

The PURC was given authority over tariff setting, and, in a critical departure from past practice, was independent of oversight from the MOME or Parliament. This was perceived as central to the reform process. PSRC recommended that tariffs be set based on established formulas for generation, transmission, and distribution. The Energy Commission was given responsibility for licensing and development of rules for the technical operation of the sector.[11]

In sum, the PSRC, drawing on the SYNEX report, advocated the rapid expansion of market principles such as unbundling VRA's generation capability from transmission and distribution, allowing open access to the transmission grid, enabling any generator to directly sell electricity to large consumers through long-term contracts, and establishing a spot market for electricity. To a substantial extent, the final reform package reflects a GOG desire for more substantial reforms based on the Chilean model rather than the limited reform implicitly suggested by the World Bank's review.

Government of Ghana "Ownership" of Reforms

That the recommendations put forward by SYNEX did shape the contours of the reform are made clear by the legal consultants hired by the PSRC, who state "proposals of the Committee are largely predicated on the system of reforms that have been implemented in Chile and Peru, and to a lesser extent in Norway and the United Kingdom" (LaBouef, Lamb, Greene, and MacRae, 1996). What explains the GOG's insistence on, and—through the use of SYNEX—development of an independent path in the electricity sector? In brief, it stems both from disenchantment with the past legacy of the sector, and a concern over how best to integrate the private sector in the future.

With regard to the past, the reform architects were motivated to more fully integrate VRA into the sector as a whole. The reformers held to the view that VRA's historically privileged position had worked against fully utilizing the benefits of cheap hydro to provide broader public benefits, particularly to serve the population rather than to serve well-paying industrial consumers. Reform afforded an opportunity to remedy this situation.

Moreover, the GOG was concerned with how best to establish a market when the sector was dominated by a single large hydropower producer. Since Chile had attempted to solve a similar problem in its own reform efforts by creating a power pool based on cost

of generation, the GOG decided to hire a Chilean consulting firm, SYNEX, with expertise on this approach.[12] It was anticipated that such an arrangement would provide opportunities for the system to use power from IPPs, and limit the extent to which the hydropower reservoir was drawn down.[13]

In addition, the GOG was wary of future arrangements with IPPs that potentially exposed the government to bearing risk through provision of sovereign guarantees. A more completely restructured sector would enable the government to require private investors to bear more complete responsibility for their revenues through, for example, long-term contracts for electricity sales. As one reformer put it, "the GOG did not want to be saddled with contingent liabilities."[14] Separating VRA into "strategic business units" and requiring competition among generators was part of this vision.

A somewhat skeptical view of the direction of reforms pursued under World Bank direction in neighboring Côte d'Ivoire was undoubtedly influential in the GOG's view of reforms. *(See Box 7.2.)* To Ghanaians, the Côte d'Ivoire experience suggested the dangers of relying on IPPs, and concerns over inviting in the private sector without clear governance arrangements. Instead of arrangements that appeared to bestow undue authority and influence on the private sector, Ghanaian reforms sought a vision of the future that married the strengths of existing public sector entities with participation as appropriate by the private sector (Opam and Turkson, 2000, p. 64).

However, this vision was by no means unanimously accepted in Ghana, and was strongly rejected by VRA. VRA executives harshly criticized the reform model as "blind copying" of the Chilean approach that was inappropriate to Ghana's circumstances. Since distribution and not generation is the biggest problem, they argued, why unbundle VRA? Instead, the focus should have been on tariff reform, which "has been a major bane to private investors."[15] Moreover, they note that a weakened VRA will be less able to perform a leadership role in development of a proposed West Africa Power Pool. However, VRA lost this battle and the MOME view prevailed.

Both sets of arguments have merit. The experience of IPPs in Asia, and the unhappy marriage of public and private sectors in Côte d'Ivoire justify caution toward an approach that invites the private sector into a country before a well-functioning market system has been established. On the other hand, the small size of the Ghanaian power sector and its weak regulatory institutions do not instill confidence that it will be able to replicate a Chilean model. That the GOG made a choice for bold reform, even at the risk of over-reaching, may have been dictated at least in part by the historical tensions over VRA's privileged status in the sector.

An additional factor may have been a sense of national pride that rejected the view that Ghana was not yet capable of developing a sophisticated market structure. Indeed, the reforms were characterized by a high degree of government "ownership." While the GOG did make use of foreign consultants, reformers are at pains to point out the extent to which the process was formulated and designed by Ghanaians.[16] For example, to build capacity at home, Ghanaian officials went on study tours to other countries to understand the options available to them. In addition, a Ghanaian national employed at the World Bank, who subsequently worked on deputation in Ghana on the reform process, played a central role in helping Ghanaian officials develop contacts and in providing expert feedback.[17]

This ownership over the process was limited to government officials. There is no evidence of engagement with civil society groups, either to solicit opinions, advice, or even to share information. Some academics did serve on task forces related to the PSRC, but their role is unclear, and they did not serve as a conduit to broader constituencies. The deliberations of the PSRC were not available to the authors in the course of conducting this study.

Did GOG ownership over reforms translate to attention to public benefits? This section examines how social concerns of access to electricity, environmental issues embedded in the reforms, and the politically sensitive tariff issue were addressed in the reform process, and with what effects.

The Social Benefits of Access to Electricity

The GOG has stated that providing the entire population with access to electricity is a long-term development priority for Ghana. In 1989, it established a National Electrification Scheme (NES) to achieve this goal by the year 2020. To what extent was this policy objective integrated into the planning of the reforms, and will the reform process work toward achieving this goal?

Reform and expansion of access were perceived as two separate and unconnected components for development of the sector.

For much of the early reform period, policies to promote universal access to electricity services proceeded on a parallel track to reform efforts. The NES pre-dates the reform efforts by over five years, and put in place a two-tier scheme for electrification. First, all district capitals (110 in total) were connected to the grid as part of a "District Capital Electrification Project." Next, under the "Self-Help Electrification Project" (SHEP), communities within 20 kilometers from the grid were invited to propose electrification projects on the condition that they procured their low-voltage poles and ensured that at least 30 percent of the households in the community were wired (Ministry of Mines and Energy, 1996; 2000). The electrification efforts were supported by a range of multilateral and bilateral development assistance organizations, led by the World Bank through the National Electrification Project initiated in 1993.

The articulated goals of the reform efforts from 1994 to 1997 do not highlight providing access to electricity. Instead, the GOG strategic framework emphasizes the need for a "reliable, efficient, and equitable supply of electricity" that serves as a basis for creating marketable assets for Ghanaian and foreign investors, ultimately leading to development of commercially viable public and private utilities (Peprah and Botchwey, 1994). However, it does not connect increasing supply to increasing access, nor is access discussed further in the details of the document. In the only mention of these issues during the reform design process, one of the reform consultants did discuss the difficulties of persuading private investors to invest in expanding distribution. The reasons given were the relatively open ended financial obligations involved, the volatility of electricity prices, and the likelihood of politicization and hence unpredictability of distribution efforts. Based on this analysis, the consultants recommended considering municipal or cooperative-based distribution systems (LaBouef, Lamb, Greene, and MacRae, 1996, p. 28). However, this proposal did not appear to have received further consideration.

Within World Bank documents, reforms are seen as parallel and separate from electrification efforts. The World Bank project document for the Takoradi thermal power plant does note the GOG's commitment to extend electricity to most of its population through the National Electrification Scheme (World Bank 1993a; 1995). However, the document does not describe how reform efforts would be integrated with the World Bank-financed National Electrification Project (a sub-component of the NES) approved only two years earlier in 1993. This shortcoming suggests that the two efforts, reform and expansion of access, were perceived as two separate and unconnected components for development of the sector. It is noteworthy that the emphasis in the pre-1993 World Bank strategy document for the electricity sector on increasing access to electricity was removed in the new strategy of 1993, which emphasized market reform and private sector involvement. Had the pre-

1993 strategy been in effect during preparation of reform efforts, World Bank task managers would almost certainly have had to report on the impacts of reform on access. Without an explicit requirement to do so, the 1995 loan document was silent on the question.

Similarly, the reform process managed through the PSRC did not explicitly reconcile the two-tier structure of electrification efforts established in 1989 with the proposed five distribution concessions created by the reforms. Instead, there was a vague expectation that electrification efforts would be "dove-tailed with reforms at a later date."[18] One observer close to the process argues that social concerns such as access to electricity have received only lip service that has not been matched by action.[19]

A subsequent Statement of Power Sector Development Policy from the GOG in 1999, which is intended to reiterate its commitment to reforms, is far more explicit about the connection between the reform and efforts at expanding access. The document invokes the government's Vision 2020 plan, which aims at full electrification and reiterates that "the overall goal of the GOG is to ensure that Ghanaians have universal access to electricity" (Ministry of Mines and Energy, 1999). Moreover, the document states that "...to ensure that the GOG's goal for universal access to electricity can be cost effectively pursued even after the privatization of the distribution utilities, a two-tier structure will be introduced alongside the establishment of the proposed distribution areas."

Under this approach, within each concession, the first tier under the NES, comprised of district capitals, would be classified as "commercial electricity zones" in which distribution licensees would have an obligation to provide service on demand. The second tier of smaller communities would be classified as "SHEP electrification zones," where it is not feasible to require an obligation to serve, but where licensees would provide services under operations and maintenance contracts on behalf of the GOG. Over time, the government hopes to increase the private sector's role in delivery of

distribution and retail service, thereby freeing public sector resources to support expansion of coverage in SHEP areas (Ministry of Mines and Energy, 1999, p. 8). The GOG also proposed establishing a "National Electrification Fund Board" with responsibility for mobilizing funds from domestic sources (such as levies), donors, and the private sector.

The belated attempt to integrate efforts to expand access and broader structural reforms may well have both created obstacles and missed opportunities. Obstacles include the challenge of integrating a private-sector-led vision of distribution concessions based on privatizating of existing assets with a continuing effort to deploy public assets in electrification efforts. Moreover, by separating publicly funded grid extension from the rest of the reform program, the GOG may have lost an opportunity to complement its electrification program by creatively harnessing private sector capabilities to serve rural areas through off-grid distributed technologies. Yet, at the time there were few examples that Ghana could draw on of public efforts to support expansion of access by facilitating private sector entry. In recent years, efforts by Chile, Morocco *(see Box 2.6 in Chapter 2)*, and Argentina *(see Chapter 3)* have shown the way toward expansion of access that is not entirely reliant on public grid expansion.

Promoting Environmental Benefits through Reform

The introduction of private sector actors and market competition held considerable potential to shift the environmental profile of the electricity sector in Ghana. Was this connection between sector reform and environment recognized and internalized in reform dialogue, planning, and implementation?

A review of reform documentation shows little or no explicit attention to environmental considerations in reform processes. The GOG's strategic framework does not recognize reform as an opportunity to provide incentives for a more environmentally sustainable electricity sector, with the exception of attention to energy efficiency (Peprah and Botchwey

1994, p. 13). Moreover, environmental issues were not discussed during PSRC deliberations.[20] Indeed, written and verbal accounts of the process suggest that environmental impact was viewed only in a damage limitation context. PSRC members assumed that environmental outcomes would be tackled by the appropriate regulatory body, such as Ghana's Environmental Protection Agency.[21] In another example, the World Bank's 1995 loan document only examines environmental questions related to siting of the proposed Takoradi power plant and related questions of obtaining permits (World Bank, 1995). Moreover, the terms of reference agreed to by the World Bank and the GOG for the PSRC makes no mention of integrating environmental objectives into the mandate of the Committee (World Bank, 1995, Annex 3.2). Finally, no civil society organizations emerged to break this silence on environmental concerns.

A significant exception to the general lack of focus on environmental factors is an explicit commitment by the GOG to promote energy efficiency and conservation as part of reform efforts. *(See Box 7.5.)* Ghana has a history of promoting energy efficiency going back to 1985, when an energy fund was established to promote efficiency (and renewable energy sources) as a means of coping with drought-induced shortages. The medium-term plan issued for Ghana's "Vision 2020" highlights increased conservation and promotion of demand side management as an objective for the sector (Government of Ghana, 1997). Moreover, tariff increases have heightened interest in conservation among industrial and

<table>
<tr><td>BOX 7.5</td><td>POTENTIAL CONSUMER SAVINGS FROM ENERGY EFFICIENCY</td></tr>
</table>

Refrigerator Standards – $50 million
A/C standards – $8 million
Lighting standards – $6 million

Source: Lawrence Berkeley National Laboratory. 1999. "Ghana Residential Energy Use and Appliance Ownership Survey." Berkeley, CA.

commercial consumers (Peprah and Botchwey, 1994). It is significant that energy efficiency measures also provide political benefits by buffering the impact of tariff increases, thereby making them more palatable. Existing studies suggest the potential cost savings are considerable. Perhaps for this reason, the GOG requested that a demand side management (DSM) component be included in the 1995 IDA credit for the Takoradi project.[22]

In partnership with the Ghanaian "Private Enterprises Foundation," the GOG established an "Energy Foundation" in 1997 to promote sustainable development of energy resources; educate consumers; advocate policies for sustainable development and enhanced customer service; strengthen private sector efforts at energy efficiency and renewable energy; and undertake related research and development (Energy Foundation, 1999). To date, the Energy Foundation has had some success, convincing industry in particular of the viability of an energy efficiency strategy. For example, by 1999 the Energy Foundation had installed capacitors in more than 100 industrial, commercial, and public organizations, freeing about 20 megavolt amperes of reactive power into the national grid. The foundation has also embarked on a sustained public education campaign to raise consumer awareness about the benefits of energy efficiency. A survey conducted by the Energy Foundation revealed that the program has been quite effective, with about 57 percent of the population attributing savings made in their electricity bills to the Energy Foundation's campaigns (Energy Foundation, 1999).

A second exception to the general lack of attention to environmental concerns was the provision for an Embedded Generation Facility (EGF).[23] An EGF was defined as a power generation facility with capacity of less than 50 megawatts, and whose output is distributed and retailed locally without use of the high voltage transmission grid (Electricity Regulation, 2000). During the electricity crisis of 1998, however, the EGF became for a time a vehicle for rapid introduction of small-scale diesel generators to meet the supply shortfall. According to the GOG's Statement of Power Sector Development Policy released in 1999, promoting the development of renewables was a key reform objective and the EGF was the vehicle for this objective.

For at least two reasons, the EGF has come to be seen by both energy practitioners in Ghana and by donors as a potential tool for promoting of renewable energy technologies, particularly wind and solar. First, by creating a legal space for private generation and sale of electricity on a small scale, the EGF will open a door to all small-scale generation technologies in Ghana. Second, the EGF will provide a further boost to renewable energy technologies through rate-setting guidelines. Specifically, the PURC will be required to set EGF bulk tariffs based on avoided costs plus the transmission service charges saved by generating electricity close to the end user. Consequently, renewable technologies are likely to be favored by distribution licensees as a cheaper alternative to meeting their service obligations than relying on centrally generated electricity. Building on this opportunity, the Danish bilateral donor agency, DANIDA, has funded a Renewable Energy Development Project to identify and remove further barriers to the promotion of renewable energy technologies in Ghana. The Danish government is also funding development of a Strategic National Energy Plan as part of a broader program of bilateral cooperation on energy policy (DANIDA, 1999).

In the long run, the scope to proactively address environmental concerns within a broader policy and planning authority rests with the Energy Commission, which was established in the course of reforms. There is some early indication that, just as cheap hydropower is currently used to subsidize more expensive thermal power, in the future the benefits of cheap hydro may be used to promote renewable energy technologies.[24] This proposed approach is an indicator that the commission is open to a range of ideas for future planning in the sector.

Tariff Setting and Good Governance

Tariff setting has long been a controversial issue in Ghana's electricity sector. Tariffs keeping pace with costs is an important component of financial viability for the electricity sector. On the other hand, high tariffs impose a considerable burden, particularly if they apply to poorer sections of the population. The resultant discontent is also politically dangerous for governments. The GOG's reluctance to consistently raise tariffs to keep up with costs was one of the contentious issues in negotiations over the Takoradi loan.[25] A detailed tariff formula was eventually negotiated with the GOG and included by the Bank as a condition for the loan.

The political dangers of tariff increases were alarmingly displayed in May 1997, prior to implementation of the PSRC recommendations, when the Ministry of Mines and Energy attempted to raise tariffs by approximately 300 percent in order to meet World Bank conditions. The tariff increase led to a national uproar. The president personally intervened to roll back the increase, and summoned Parliament from its recess to approve the then-pending draft legislation to establish an independent regulatory agency. The PURC became law shortly thereafter, in October 1997.

Interestingly, the PURC managed to successfully pass an equivalently large 300 percent increase during 1998 without arousing nearly the same degree of popular discontent. PURC staffers have attributed this success to their concerted public outreach and dialogue—including workshops, public forums, and media campaigns—prior to raising tariffs.[26] According to staff, central to their argument

was persuading the population that the tariff hikes would be used responsibly, for purposes such as increasing access to electricity for the rural population. While this campaign may well be part of the story, electricity sector officials also widely recognize that a new power crisis caused by yet another drought in 1997 was also instrumental in convincing the population of the need for sacrifice in the long-term interest of the sector (Peel, 1998).

Despite these efforts, battles over tariffs continue. The tariffs for the dominant residential consumers are still not close to economic levels (even after the May 2001 tariff increment averaging about 95 percent for all categories of end user),[27] and the benefits of increased tariffs are normally eroded by devaluation of the national currency. Consequently, the debate continues over whether tariffs should be determined entirely based on the financial viability of the utilities, and hence by technical and economic criteria, or whether some measure of social impacts should also be considered. In its comments, the World Bank has stressed the need to ensure that the PURC is governed by rules that ensure not only consumer rights, but also those of suppliers (Tomlinson, 1997). For its part, the PURC has felt itself under pressure to raise tariffs, notably from the VRA and from international donor agencies, but it has resisted on the grounds that utilities have made insufficient efforts to increase their efficiency, reduce losses, and improve quality of service.[28] The commission has decided, through a "transitional plan," to ultimately achieve economic tariffs within a three-year period for all categories of consumers. (*See Box 7.6.*) The PURC's objective, among other things, is to introduce some amount of gradualism in the attainment of economic electricity rates to minimize the impact of tariff increments on all classes of customers (PURC, 2000).

The PURC arguably has "endeavored to steer a middle course," recognizing that its decisions "cannot realistically be insulated from the macroeconomic, as well as the socioeconomic environment of the country" (Opam, 1999). To ensure that it steers this course correctly, the PURC has attempted to incorporate several principles of good governance in its structure and functioning. These include institutional representation of industry, labor, and domestic consumers on the Commission; transparent guidelines for tariff setting; publication of tariff decisions; and mechanisms for public hearings and representation before the PURC (Opam, 1999). Early indications suggest that these mechanisms are functioning well.

CONCLUSION

Ghana was forced to consider reforming its sector by a combination of demand shortfall—in part due to drought—and the drying up of its traditional source of financing, the World Bank. There is little doubt that the World Bank was instrumental in urging the government to seriously consider a program of reform. All sides are agreed that without the threat of a cut-off of World Bank funds, the government would not have undertaken substantial reforms.

However, in designing the program, the government forged its own path, and one quite distinct from World Bank recommendations. Instead of pursuing limited reforms and relying on management contracts to improve performance, the government embraced far-reaching reform, perhaps to more completely integrate the giant VRA into the sector.

Did government ownership translate to an emphasis on a public benefits agenda? Certainly, part of the motivation for integrating VRA into the sector was to spread the benefits of VRA's cheap hydropower more broadly through the population. But given the slow pace of the reform, whether the public will benefit from this approach is as yet undetermined. While the government had a long-term commitment to expand access to electricity, for much of the reform process access and reform proceeded on parallel tracks. In its support for both reforms and for electrification programs, the World Bank also pursued both as separate projects. Only in 1999 did the government make explicit attempts to relate access to the reform process.

On the environmental front, the government has proactively pursued energy efficiency measures, partly because it enables them to blunt the effect of tariff increases. With funding from the Energy Fund during the late 1980s and the more recent assistance of donor agencies, particularly the Danish government, it has also pursued an active program of promoting small-scale renewable energy sources. Indeed, the promotion of renewable energy by private developers has been made possible mainly by the reform program's initiative in creating a legal spacc of private generation and sale of electricity.

The Energy Foundation has convinced industry of the viability of an energy efficiency strategy.

Despite these advances, electricity sector reform in Ghana has proceeded slowly, in no small part because of the difficulties of making decisions that would impact harshly on the general populace. At root is a problem of tariffs that are low when viewed through the eyes of utilities or investors seeking to recover their costs, but very high when viewed through the eyes of Ghanaian consumers. The tariff problem is worsened by periodic devaluations of the currency. Indeed, the Ghanaian problem with tariffs in a climate of macroeconomic uncertainty may be a nearly intractable obstacle to applying a profit-making model to a sector in a poor country.

NOTES

1. The authors wish to acknowledge the considerable contribution of Abeeku Brew-Hammond and Martin Bawa Amadu to this chapter, which draws on Ishmael Edjekumhene, Martin Bawa Amada, and Abeeku Bew-Hammond, "Power Sector Reform in Ghana: The Untold Story," unpublished paper submitted to the World Resources Institute. February 2001, available at http://www.wri.org/governance/iffepowercases.html.

2. Interview with World Bank official, September 22, 2000. All interviews for this chapter were conducted on a not-for-attribution basis. Consequently, interviewees are identified only by their institutional affiliation.

3. VALCO's consumption decreased to 39 percent of generation by the late 1990s due to diminished hydro capacity following a drought season. By 1996, industrial consumers accounted for 68 percent of sales, the residential sector 26 percent, and commercial consumers 5-6 percent (Opam and Turkson, 2000).

4. Since 1994, powcr flows have reversed, with Ghana importing power from Côte d'Ivoire (Opam and Turkson, 2000).

5. Interview with former power sector official, September 22, 2000.

6. VRA had a contractual agreement with VALCO to supply the latter with not less than 2,760 gigawatt-hour of power annually. VALCO suffered a 95 percent cut in supply during the 1982-83 drought (Brew-Hammond, 1994). An attempt by VRA to curtail power supply to VALCO during the 1994 crisis resulted in the latter taking VRA to court for breach of contract (Daily Graphic, 1994). In February 2002, VALCO took the GOG to court because the latter asked it to shut down two of its pot lines. The GOG said this would enable the system to have a reserve capacity of 150 megawatt. The court ordered the parties to negotiate a compromise, which led to the shutting down of only one pot line instead of two. The GOG negotiating team has also indicated that VALCO should be prepared to pay higher tariffs. Currently VALCO pays 1.7 cents per kilowatt-hour compared to 3.99 cents per kilowatt-hour paid by residential consumers. (*Daily Graphic,* 2002a; 2002b).

7. Interview with senior GOG official, December 6, 2000.

8. While this was a majority view at the time, it was by no means the only view within the Bank. Indeed, there were other voices from within, as we discuss later, that were in favor of a more comprehensive market-based reform.

9. This impression is based on interviews with members of the Power Sector Reform Committee (PSRC), December 6, 2000, and January 11, 2001.

10. This description is drawn from Opam and Turkson (2000).

11. For example, generation tariffs for the regulated market were set on the basis of short run marginal cost. Distribution tariffs would be based on long run marginal cost and distribution value added, the latter computed using an average reference level to provide an incentive for efficiency improvements.

12. Personal communication with World Bank staff, March 21, 2002.

13. Large swings in reservoir levels brought environmental problems and difficulties for the communities that lived along the reservoir.

14. Personal communication with World Bank staff, January 2000.

15. Interview with VRA official, September 12, 2000.

16. Interview with members of the PSRC, September 13, 2000, and September 15, 2000.

17. Interview with PSRC member, December 6, 2000.

18. Interview with PSRC member, December 6, 2000.

19. Interview with member of the Energy Commission, September 14, 2000.

20. Interview with PSRC member, September 13, 2000.

21. Interview with PSRC member, September 13, 2000.

22. The World Bank provided an IDA credit of $4 million toward a total cost of $8.5 million for this component (World Bank, 1995).

23. The Energy Commission Act of 1997 mandated the preparation of electricity regulations, which were prepared in 2000. These regulations explicitly mention the EGF.

24. Interview with member of the PSRC, September 15, 2000.

25. Interview with reform official, September 22, 2000.

26. Interview with PURC member, December 6, 2000, and reform official, September 22, 2000.

27. Currently, residential consumers pay 3.99 cents per kilowatt-hour as opposed to the indicative economic rate of 7.43 cents per kilowatt-hour. Meanwhile, in their proposal for review of tariffs submitted to the PURC in March 2001, the utilities asked for 10.05 cents per kilowatt-hour.

28. Interview with PURC member, September 15, 2000.

REFERENCES

Berenschot, Moret, and Bosboom. 1985. "Study of the Public Enterprise Sector in Ghana, Final Report Vol. 1-A, The SOE Sector in Ghana." Management Consultancy Report.

Brew-Hammond, A. 1994. "Finance and Markets for Electric Power Development in Ghana." *Science in Africa: Energy for Development Beyond 2000.* Paper presented at a conference organized by the AAS sub-Saharan Africa Program, Washington, D.C.

Daily Graphic. 1994. "VALCO Takes VRA to Court." No. 13613 (September 6):1.

———. 2002a. "Pay More for Power Supply, Says Government Team." No. 148439 (February 4):1.

———. 2002b. "Power Crisis: VALCO Ordered to Close 2 Pot Lines." No. 148453 (February 20):1.

DANIDA. 1999. "Energy Sector Program Support for Ghana: Elaboration of a Strategic National Energy Plan, 2000-2020."

Edjekumhene, I. 2000. "Privatization in Ghana." Unpublished MSc. Thesis. University of Hull. United Kingdom.

Energy Foundation. 1999. *Annual Report.* Accra.

Government of Ghana. 1997. "Ghana-Vision 2020: The First Medium-term Development Plan (1997-2000)." National Development Planning Commission. Government of Ghana. (July).

Kapur, I., Hadjimichael, M.T., Hilber P., and Szymczak, P. 1996. "Ghana: Adjustment and Growth–1983-1991." Occasional Paper No. 86 (September). Washington, D.C.: International Monetary Fund.

Ministry of Mines and Energy. 1996. "Energy Sector Development Programme (1996-2000)." Accra.

———. 1999. "Statement of Power Sector Development Policy." (April) Accra.

———. 2000. *Electricity Regulation.* Accra.

Opam, M. 1999. "Post Reform Issues: The Experience of Ghana." Paper presented at Power Sector Reform and Regulatory Strategy conference, Kampala, Uganda, October 20-21.

Opam, M. and J.K. Turkson. 2000. "Power Sector Restructuring in Ghana." In *Power Sector Reform in sub-Saharan Africa.* J.K. Turkson ed. London: Macmillan Press Ltd.

Partiff, T.W. 1995."Adjustment for Stability or Growth? Ghana and Gambia." *Review of African Political Economy.* No. 63:55-72.

Peel, Quentin. 1998. "Return to the IMF Embrace." *Financial Times.* Survey on Ghana. (June 22).

Peprah, Kwame, and Kwesi Botchwey. 1994. "Strategic Framework for Power Sector Development Policy." Letter to Joanne Salop, Acting Director, Western Africa Operations Division, World Bank. (March 23).

Public Utilities Regulatory Commission. 2000 "Ghana's Energy Pricing." Paper presented at 5th Kumasi International College on Energy conference, Kumasi, Ghana, March 27-April 2.

Tomlinson, Mark D. 1997. "Public Utilities Regulatory Commission Bill." Memo to Mr. Fred Ohene-Kena, Minister, Ministry of Mines and Energy. (September 26).

World Bank. 1993a. "Staff Appraisal Report, Republic of Ghana, National Electrification Project." Washington, D.C.

World Bank. 1993b. *The World Bank's Role in the Electric Power Sector.* Washington, D.C.

World Bank. 1995. "Staff Appraisal Report, Republic of Ghana, Thermal Power Project." Washington, D.C.

8

SOUTH AFRICA

ELECTRICITY REFORM WITH A HUMAN FACE?

Julia Philpott
Alix Clark[1]

INTRODUCTION

South Africa's unique colonial history, apartheid legacy, and ongoing transition to democratic governance drive the country's determination to attain its development objectives. Embedded in that determination is a broad social and environmental public benefits agenda—that is, a sustainable development agenda. Public benefits include energy access, grid and off-grid electrification, black economic empowerment, renewable energy, and energy efficiency. South Africa's determination is not an automatic guarantee that public benefits can or will be incorporated into the electricity sector reform process. Nor is it a guarantee that the value of public benefits will materialize and be tangible in the daily lives of South Africans. The reform process is nascent in South Africa in comparison to many other countries, and there is an ongoing debate over the electricity sector's ultimate governance structure. This debate reflects a broader, national debate about how South Africa will attain its sustainable development agenda. At its core is a choice between two very different approaches: centralized/government-directed versus decentralized/market-driven. Nevertheless, that the government has begun framing reform around development objectives, such as black economic empowerment and electrification, distinguishes South Africa's process as unique among others. That the process is still in its early stages presents opportunities to contour it further to have a human face—one whose characteristics include a suite of public benefits—representative of all South Africans.

BOX 8.1 | PROFILE OF THE ELECTRICITY SECTOR IN SOUTH AFRICA

Population (2001): 40.3 million [1]

Percentage with access to electricity (1999)[2]: 66%
    Rural: 46%   *Urban:* 79%

Installed electricity generation capacity (1999)[3]:
  48 gigawatts (1.19% of total world capacity)
    Coal: 91%
    Nuclear: 7%
    Other: 2%

CO_2 emissions from electricity and heat as a share of national emissions (1999)[4]: 61%

Notes:

1. World Resources Institute. 2000. *People and Ecosystems: The Fraying Web of Life*. Washington, D.C.: World Resources Institute.

2. National Electricity Regulator. 2000. South Africa.

3. www.eia.doe.gov/pub/international/ieapdf/t06_04.pdf (February 6, 2002).

4. Computed by WRI using International Energy Agency (IEA) data. IEA, 2001. *CO2 Emissions from Fossil Fuel Combustion*. Paris: OECD.

Apartheid Era Influences

During the apartheid era, South Africa's national energy policy called for the nation to be self-sufficient through its own domestic energy production.[2] In accordance with national policy, South Africa's state-owned utility, Eskom, generated almost all of the country's electricity. Eskom fueled its central-station power plants with cheap coal from South Africa's abundant reserves.[3] For a profile of the sector, see Box 8.1.

Throughout the 1970s, armed conflict escalated within South Africa due to the ruling National Party's (NP) apartheid policies. By the early 1980s, the international community imposed economic sanctions on South Africa—including an oil embargo—to signal the country's deepening status worldwide as a pariah state. At the same time, the NP-led government continued to invest millions of Rand in the electricity sector in the form of new coal-fired, central-station power plants. Despite insufficient electricity demand to warrant new power plants, Eskom also spent millions of Rand to build the Koeberg nuclear power plant near Cape Town.

Access to electricity became a symbol of social and economic equity in South Africa.

Domestic and international pressures eventually had an impact on South Africa's top political leadership. In 1990, former President and NP leader F.W. de Klerk released African National Congress (ANC) leader Nelson Mandela from prison. In 1994, Mandela ran in South Africa's presidential election and won with the Reconstruction and Development Programme (RDP) as his political platform.[4] At the start of Mandela's presidency, 40 percent of the population had electricity (Wamukonya, 2001). In urban black municipalities, approximately two thirds of the population had electricity. In rural black municipalities, the majority of the population had no electricity. Those who had were subject to unreliable, poor-quality service from a decaying electricity infrastructure.

Development and Energy: Priorities for a New South Africa

The RDP established South Africa's social and economic development priorities: economic growth, job creation, and access to services that meet basic human needs among others—including energy. With regard to the electricity sector, the RDP mandated that during the 1994–2000 period the government had the responsibility to oversee 2.5 million household connections to the electricity grid. The government via Eskom surpassed its own electrification target, which was set at 350,000 connections per year. The percentage of the population with access to electricity rose from 40 percent in 1994 to 66 percent in 2002 (Wamukonya, 2001). By 2002, 79 percent of the population in urban areas and 46 percent in rural areas had access to electricity (NER, 2000). Eskom financed the electrification program and paid for it through a cross-subsidy tariff.[5]

The Department of Minerals and Energy (DME), under the current leadership of Minister Phumzile Mlambo-Ncguka, sets the vision for South Africa's national energy policy broadly and by sector. Under DME's guidance, the electricity sector's policy focus on universal electrification and greater energy access for the poor increased at apartheid's end. At the same time, the policy emphasis on domestic energy self-sufficiency decreased. Universal access to electricity became a symbol in the "new" South Africa of social and economic equity. In 1998, DME published its White Paper on the Energy Policy of the Republic of South Africa 1998.[6] The document spelled out the government's energy policy and its major objectives: (1) increasing access to affordable energy services; (2) improving energy governance; (3) stimulating economic development; (4) managing energy-related environmental impacts; and (5) securing energy supply through a diversity of sources. In a speech at the Africa Energy Forum, the DME minister gave the following description of reform's relationship to South Africa's social and economic development:

> "The energy sector is being restructured in order to ensure that we reduce the cost of energy, improve economic efficiency, attract local and foreign direct investment, diversify energy resources for environmental reasons, and ensure security of energy supply. This will ensure the delicate balance between State's imperative to spur on economic growth and its social responsibilities...[In] order to limit the expected upward pressure on electricity prices due to Eskom's new dividend and tax payments status, there is a need to build new generating capacity, to encompass environmental considerations, and to reform the [sector]. Government will ensure that these price increases will be kept as low as possible, so that South Africa maintains its competitiveness, cross subsidies for the poor, and free basic services to bring relief to the poor" (Mlambo-Ncguka, 2001).[7]

CONTEXT FOR SOUTH AFRICA'S ELECTRICITY SECTOR REFORM

South Africa is still experiencing a historic transition from apartheid to democratic governance. Governance expresses the notion that, as part of the national political process, the government determines—with broad stakeholder participation—the forms of utility ownership and sector management in a transparent, open, and participatory manner. The government makes this determination based on its social and economic policies and ideological leanings (Steyn, 1995). The debate in South Africa over the electricity sector's ultimate governance structure reflects in some measure a broader, ongoing national debate about how the country will attain its social and economic development objectives (Galen, 1998). The following is a brief description of Eskom, the role of municipal governments in the reform process, and an overview of the opposing positions of national government and labor interests.

Eskom: South Africa's State-owned Utility

Eskom is South Africa's state-owned electric utility that has been a vertically integrated monopoly since its inception.[8] It currently owns and operates 24 of the country's 53 central-station power plants. Municipalities own 22 power plants and private interests own the remaining 7. South Africa has a total licensed capacity of 48,124 megawatts.[9] Eskom holds a licensed capacity of 44,852 megawatts (93 percent of the total), and 368 municipalities hold 2,436 megawatts (5 percent). A handful of private interests hold 836 megawatts (2 percent) of licensed capacity (NER, 1999). In 1999, South Africa's national peak demand was 29,398 megawatts, and its total surplus electricity generation capacity was 18,726 megawatts (NER, 2000). Eskom also owns and operates the country's transmission system, which connects the power stations to large urban and industrial areas within South Africa and to its neighboring countries.

Municipal Governments: Key Actors in Electricity Distribution

In South Africa, municipalities are responsible by law for providing infrastructure and public access to such services as water, sanitation, telecommunications, and electricity. As distributors of electricity whose function is to sell the commodity to households, businesses, and other customers they have constituted historically an influential group. The government's political and financial commitments to universal electricity access increased their standing as a large and influential political constituency group. While few municipal distributors generate the electricity they sell, the majority purchase it at wholesale prices from Eskom. In 1999, there were 6.6 million electricity customers in South Africa.

Municipal distributors and private interests serviced 49 percent of this group, and sold 41 percent of the total electricity consumed by them. Eskom serviced 50 percent of these customers, and sold 59 percent of the electricity they consumed (NER, 2000).

Government Favors a Decentralized/Market-driven Development Approach

On a macroeconomic level, the government appears to support a market-driven approach. For instance, the market-oriented approach in the Growth, Employment, and Redistribution (GEAR) program has complemented—if not superseded—the RDP as South Africa's macroeconomic strategy.[10] In another example, the government announced plans in 1994 to restructure South Africa's four largest state-owned enterprises (SOEs), including Eskom. The Department of Public Enterprise (DPE), headed at present by Minister Jeff Radebe, guides the SOE restructuring process. In 2000, DPE articulated its vision for the ownership of South Africa's SOEs—which included Eskom—in the framework document An Accelerated Agenda Towards the Restructuring of State Owned Enterprises: Policy Framework (DPE, 2000). The following statement by Minister Radebe captures the rationale for SOE restructuring:

> "[Our] interest in restructuring as an instrument of public policy emanates from the very real need to ensure that our people as a whole live better, engage with one another with greater humanity and justice, are exposed to the opportunities and chances that come with greater social wealth, so that they can take their place proudly as citizens in every sense of the word, of South Africa, of Africa and of the world. This is our emancipating vision and one that drives our thinking. As such, restructuring is not driven by cold dogma or rampant ideology. As a tool it has certain dimensions, but many applications. It is our task, nay, responsibility, to work out the particular job to be done for the greater benefit of the people of our country" (Radebe, 2000).

Labor Interests Favor a Centralized/Government-driven Development Approach

Not all stakeholders agree the electricity sector needs reform. Despite reform's focus on black economic empowerment, some labor interests do not agree that Eskom's restructuring, and SOE restructuring in general, will automatically lead to an improvement in peoples' lives and livelihoods. The Eskom Conversion Bill, put forth by DPE last year, became a rallying point for labor interest's opposition to SOE restructuring.[11] The Congress on South African Trade Unions (COSATU)—one of South Africa's largest and most influential trade unions—called for DPE to withdraw the bill immediately. COSATU's greatest concern about the restructuring was potential job losses. In a May 2001 press release, the trade union expressed its discontent with the process and announced the severance of its relations with DPE.[12] On August 29 and 30, 2001, COSATU called a political strike—it was one of the first strikes aimed explicitly at the government's SOE restructuring policy (Winkler and Mavhungu, 2001). Privatization was one of the focal points for the strike.[13]

The focus on empowerment as part of a reform process finds echoes in other countries. Box 8.2 describes an effort in Bolivia at privatization through broad capitalization of a state-owned utility aimed at serving social equity goals.

ELECTRICITY SECTOR REFORM DRIVERS IN SOUTH AFRICA

Reform in South Africa is really two processes unfolding simultaneously and separately. First, there is the electricity distribution industry (EDI) reform

BOLIVIA'S CAPITALIZATION PROGRAM: PRIVATIZATION THAT SERVES SOCIAL EQUITY GOALS

The idea for a modified privatization program that would serve larger equity and social development goals originated with former Bolivian President Sanchez de Lozada, and drew from his "Plan for All" political platform. This "capitalization plan" resulted from a compromise among important interest groups, including members of the president's social democratic political party (MNR), labor unions (Central Obrera Boliviana), constituencies concerned with the sale of strategic assets to foreign investors (Comité de Defensa del Patrimonio Nacional), Bolivian business networks, policymakers, and donors advancing market-oriented proposals.

The capitalization plan, which included but was not limited to the electricity sector, involved selling 50 percent ownership of state-owned enterprises in strategic sectors to private investors in a transparent bidding process. The other 50 percent was transferred in the form of shares to Bolivia's working-age population. In the capitalization scheme, the price paid by the private buyer for the state asset is reinvested in the enterprise to improve its performance and upgrade infrastructure. The bidding process requires companies to submit investment plans, and bids are evaluated in part on the strength of the investment plan.

The capitalization program combined several structural reform efforts, including a wholesale reform of the social security and pension system. The shares transferred to citizens at large were distributed in two ways. About 70 percent of shares were credited to individual retirement accounts opened under Bolivia's pay-as-you-go pension fund system. This new system was created as part of a wholesale transformation of Bolivia's social security programs at the same time that the capitalization process gained momentum (1997). Private pension fund companies manage pension accounts, and account holders are free to trade the shares transferred to them. The remaining 30 percent of shares were deposited into a fund (the Capitalization Fund) that is also managed by private pension companies. The earnings generated by this fund are distributed as annuities, called the "bonosol." Beginning December 31, 1995, Bolivians between the ages of 21 and 50 are eligible to receive the *bonosol* upon retirement.

The compromise that resulted in the capitalization proposal acknowledged the interests of powerful national constituencies, but it also reflected Sanchez de Lozada's interest in the preservation of Bolivia's collective social traditions, particularly traditional indigenous forms of economic organization. Capitalization was not exclusive to the electricity sector. Other reforms strictly tied to the sector received considerable support from the World Bank. In 1993, de Lozada's government solicited support from the World Bank's Energy Sector Management and Assistance Programme (ESMAP), and subsequently received technical assistance from the Bank (a total of $156 million from IDA). Consultants from Chile, Argentina, and Peru were hired with these funds to help define a new structure for the electricity sector. According to ESMAP, there was strong Bolivian leadership in the design of electricity reforms.

Sources:

Griffin, Paul. 1999. "Privatization in Bolivia." Cadwalader Newsletter on *Global Project Finance & Privatization.* (Summer).

Kritzer, Barbara E. No date. "Social Security Privatization in Latin America." Policy paper produced for the U.S. Social Security Administration.

Rufin, Carlos. No date. "Institutional Change in the Electricity Industry: A Comparison of Four Latin American Cases." Unpublished manuscript. John F. Kennedy School of Government. Harvard University.

World Bank. 2000. "Introducing Competition in the Electricity Supply Industry in Developing Countries: Lessons from Bolivia." World Bank ESMAP Report No. 233/00. Washington, D.C.: ESMAP. (August).

process, which began in 1992. The EDI is comprised of Eskom, 368 municipalities, and a few private interests. Second, there is the electricity supply industry (ESI) reform process, which started in 1993.[14] The ESI includes Eskom, which undertakes transmission and is a large player in generation, and a handful of municipalities and private interests that generate electricity. Together, the EDI and the ESI comprise the electricity sector as a whole in South Africa. The government coordinates both reform processes, but has not yet fully integrated them. The following section describes the reform drivers in the EDI and ESI.

Financial Difficulty Drives EDI Reform

Many of South Africa's municipal electricity distributors have struggled financially for years. About one third of them continue to be in severe financial straits. The problem is that some municipalities generate surplus revenue, but most operate routinely at a financial loss. By international standards, most of the distribution businesses are extremely small. In 1994, just four municipal distributors earned 50 percent of all revenues; 22 distributors earned 75 percent, and 100 distributors earned 99 percent of all revenues. The remaining 300 or more collectively earned less than 1 percent of the total electricity revenues earned by all municipal distributors (Galen, 1998). Many rely heavily on revenue from electricity sales to pay for wholesale electricity purchases from Eskom. Some also rely on cross-subsidies to provide other legally required public services.

A reason that many municipal distributors cannot operate without losing funds is because a stable economic base in the townships they serve does not exist or is insufficient. Apartheid policies prevented black municipalities (termed townships) from developing a legitimate economic base that would have created jobs, entrepreneurial opportunities, and some degree of economic stability. Compared to the white municipalities, the black municipalities usually had larger populations, most of whom were poor. Consequently, a tax base that could generate a steady stream of tax revenue could not develop (Galen,

1998).[15] At apartheid's end, the consolidation of white and black municipalities as part of the democratization process merged the economies and tax bases of areas that were formerly separate. Many municipal distributors have had to contend for years with periodic boycotts in which households refused to pay their electricity bills. Boycotts remain a means for the electorate to register periodically their disapproval of current policies.

EDI's financial difficulties produce three negative impacts. First, distributors cannot consistently provide reliable, adequate, high quality service to their customers. Second, some distributors are unable to pay Eskom for the wholesale electricity they have purchased. Third, many administrative and technical functions are duplicated across adjacent distributors in rural, urban, and industrial areas. Consequently, costs and prices in the electricity sector are unnecessarily high. These negative impacts spurred the government, Eskom, and some municipal distributors to begin grappling with EDI reform in the late 1980s.

Concerns about Future Demand for Electricity Drive ESI Reform

South Africa's vision for its electricity sector, like many other countries around the world, has gone from one of exclusive state ownership and control to one of partial private ownership and control. There are typically three main drivers motivating a move toward competition and private ownership in the electricity sector. They are desired changes in: (1) electricity prices; (2) service quality and adequacy; and (3) supply/generation capacity. Not all three drivers are relevant ESI reform motivators in South Africa. In the short term, lower electricity prices are not a strong driver, since South Africa's electricity prices are among the lowest in the world.[16] The average sales price is R0.1629 (US$ 0.0148 in 2002 prices) per kilowatt-hour (NER, 2000). Electricity service in terms of quality, reliability, and adequacy is generally sufficient for those who have it. Improved service, therefore, is not a strong driver. The driver that is strong and motivates ESI reform is the country's need for greater capacity to generate electricity in the future.[17]

In 1999, South Africa had 18,726 megawatts of surplus electricity generation capacity (NER, 1999). Sometime between the years 2007 and 2011, its surplus capacity will be depleted (Eberhard, 2001). Projections suggest that over the next 20 to 25 years the amount of new generation capacity needed might be anywhere from 24,000 megawatts to 54,000 megawatts (Galen, 1998). Some estimates suggest the level of new investment needed will be between R100 billion and R648 billion (US$9.1 billion to US$58.9 billion in 2002 prices) (Eberhard, 2001; Galen, 1998). At present, the government is the sole shareholder in Eskom, meaning it bears all the financial risk.

The sector may need US$9.1 billion to US$58.9 billion of new investment.

Two questions for South Africa become even more pressing to answer in the reform process: How and from where will the government choose to amass the investment funds needed to meet the nation's future, long-term electricity demand? The massive investment required inevitably will affect other competing development priorities—including a broad public benefits agenda—making the choice that much more difficult. The government has begun to answer these questions. It is currently restructuring Eskom in order to allow partial private investment and ownership in the generation business. As the following passage in DME's White Paper on Energy Policy (DME, 1998) conveys, the need to address the sector governance/investment question is an important driver behind ESI reform in South Africa:

> "[t]he rapid changes in the political and economic context of the electricity supply industry

worldwide in recent years raise questions about the continued ability of South Africa's monopolistic electricity industry to meet customers' electricity services needs in future."

THE DESIGN PROCESS FOR REFORM

Despite all the "right" political signals made by the government in favor of a public benefits agenda, the agenda continues to receive uneven attention in the reform process. Politicians have played a leading role in focusing the reform process on some public benefits concerns. The DME and DPE ministers, particularly the former, appear to have been responsible for ensuring that such social and economic public benefits as electrification, energy access, and black economic empowerment are on the reform agenda and discussed actively. Whether and how issues relating to end user energy efficiency and renewable energy have also had their place ensured on the reform agenda is less certain.[18]

Environmental benefits have not been prominent on the agenda. It is not obvious why this is the case, given that the RDP and *White Paper on the Energy Policy of the Republic of South Africa 1998* support an environmental focus. One reason might be South Africa's abundance of coal and its intention to use it as stated in its national energy policy. Or it could be that groups in civil society linked traditionally to environmental activism have not yet participated in the ESI reform process. Yet another reason might be that the South African government is already dealing with too many core development issues. Renewable energy's lack of prominence on the reform agenda may soon change. In a parliamentary media briefing in February 2002, the DME minister announced that her department would table at cabinet the draft *White Paper on the Energy Policy of the Republic of South Africa 1998* by mid-year. She underscored the document's importance for positioning South Africa "[to] participate and influence the agenda of the World Summit on Sustainable Development.[19] [The White Paper on the Energy Policy of the Republic of South Africa 1998] is based on the premise that renewable energy could play a small but important

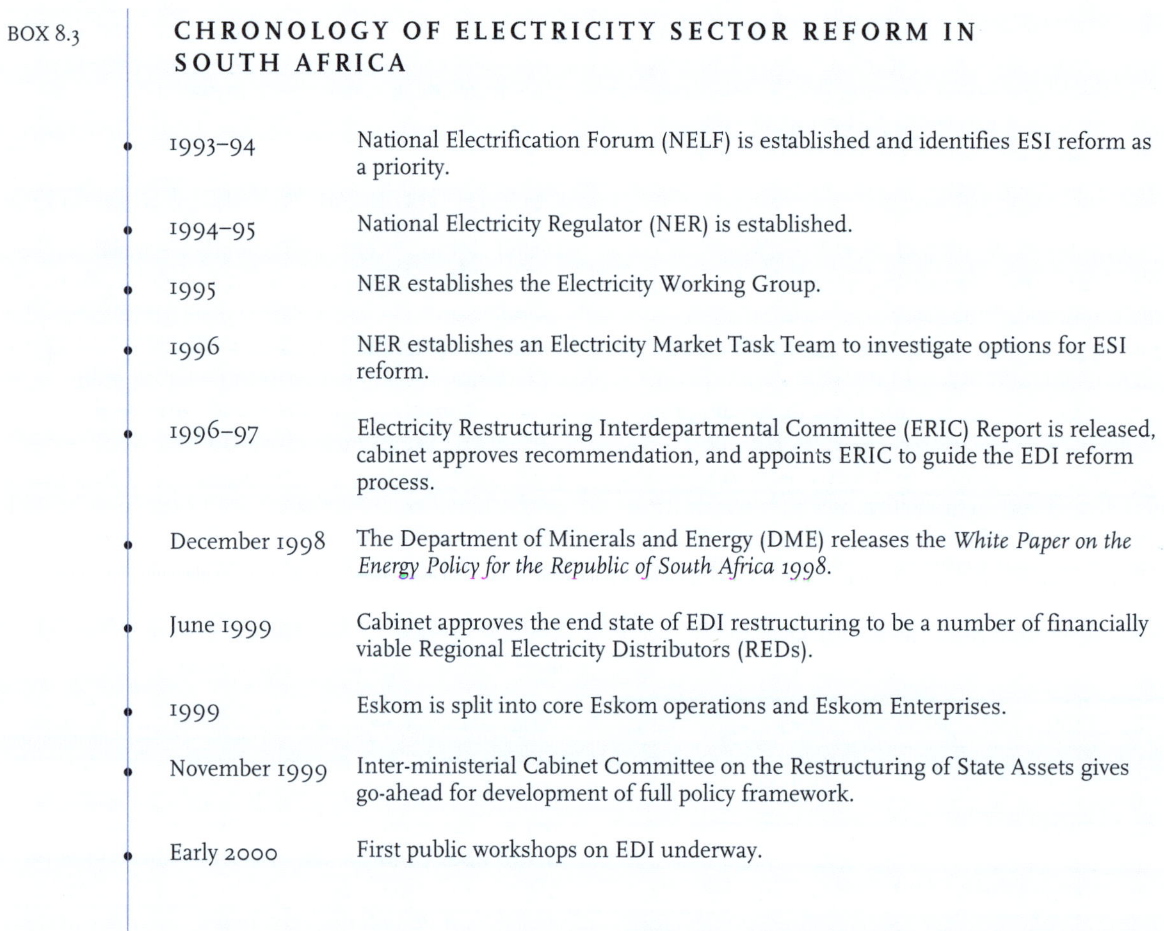

CHRONOLOGY OF ELECTRICITY SECTOR REFORM IN SOUTH AFRICA

1993–94	National Electrification Forum (NELF) is established and identifies ESI reform as a priority.
1994–95	National Electricity Regulator (NER) is established.
1995	NER establishes the Electricity Working Group.
1996	NER establishes an Electricity Market Task Team to investigate options for ESI reform.
1996–97	Electricity Restructuring Interdepartmental Committee (ERIC) Report is released, cabinet approves recommendation, and appoints ERIC to guide the EDI reform process.
December 1998	The Department of Minerals and Energy (DME) releases the *White Paper on the Energy Policy for the Republic of South Africa 1998*.
June 1999	Cabinet approves the end state of EDI restructuring to be a number of financially viable Regional Electricity Distributors (REDs).
1999	Eskom is split into core Eskom operations and Eskom Enterprises.
November 1999	Inter-ministerial Cabinet Committee on the Restructuring of State Assets gives go-ahead for development of full policy framework.
Early 2000	First public workshops on EDI underway.

role in the development of a sustainable energy system, and in ensuring a resource base for future generations" (Mlambo-Ncguka, 2002).

The following section describes the ways in which public benefits have or have not been included in the process so far. It provides a snapshot of the EDI and ESI design processes, highlights the role of the World Bank in the reform process and the role of foreign consultants in EDI reform. This section also provides examples of how Eskom's influence and municipal electricity distributor perspective have helped shape the contours of electricity sector reform and the inclusion of a public benefits agenda.

The EDI Design Process and Public Benefits

The government is seeking a structure to improve the performance and financial health of the electricity distribution industry (EDI). The current plan is to consolidate the municipal electricity distributors and Eskom into six regional electricity distribution companies, or REDs. Such consolidation allows financially weaker municipalities to benefit potentially from a merger with stronger ones. Additionally, consolidation will decrease duplication of administrative and technical functions, thereby allowing for economies of scale. Figure 8.1 is a schematic of the government's

April 2000	DME, the Department of Public Enterprises (DPE), and the World Bank conduct a workshop on ESI reform.
Mid 2000	"Stage 1 Blueprint" report for EDI reform finalised and reviewed by business and unions.
August 2000	DPE minister releases *Accelerated Agenda Towards the Restructuring of State-Owned Enterprises.*
Late 2000	DPE initiates study to make recommendations on new market structure for ESI, and cabinet requests DME to submit a strategy for ESI reform.
Early 2001	On directive of DME minister, a team of international and local electricity sector specialists reviews the "Stage 1 Blueprint" for EDI reform.
May 2001	Parliament enacts the Eskom Conversion Act.
2001	Cabinet adopts the "Stage 1 Blueprint" for EDI reform.
May 2001	On directive of the DME minister, a team of international and local electricity specialists reviews the Norwegian-supported study on ESI reform.
Mid 2001	NER grants new licenses to independent power producers to build new capacity.
Late 2001	National EDI Holdings Company established.
2001	Eskom generation units are clustered into competing businesses.
2002	REDs established.

proposed, near-term structure for South Africa's electricity sector as a whole. The section following provides examples of different viewpoints and positions of different EDI stakeholders.

Foreign Consultants: PricewaterhouseCoopers

In early 2000, the Department of Minerals and Energy (DME) hired the U.K.-based consulting firm PricewaterhouseCoopers (PwC) as its technical advisor on EDI design and strategy. DME proceeded to organize and work with PwC on a series of consultative meetings with business and labor interests to discuss EDI reform. By June 2000, PwC had published a series of six working papers for DME that outlined a possible "blueprint" for EDI reform.[20] The six papers focus on the definition of the REDs; ownership, governance and legal status; asset valuation and transfer; regulation and commercial arrangements; tariffs and financial transactions; and organization and human resources. These discussion documents made little mention of the provision of public benefits—such as universal grid and off-grid electrification, energy access, energy efficiency, renewable energy, and black economic empowerment.

In late June, PwC published a seventh working paper entitled "Working Paper 7: Consolidated Emerging Views" (PwC, 2000). This paper "consolidates the key elements of the six working papers and, in some areas, indicates how [the] thinking has developed since the workshops" (PwC, 2000). The government relied on and used much of the substantive content of this PwC paper as the basis for its "Stage 1 Blueprint" report for EDI reform.[21] In November 2000, DME submitted to cabinet—after stakeholder scrutiny and comment—the "Stage 1 Blueprint" report. After reviewing the report, the cabinet remanded a series of questions to DME primarily about whether, how, and to what degree the blueprint addressed the needs of poor South Africans (Clark, 2001). At the behest of the government, the revised "Stage 1 Blueprint" described South Africa's electrification program as an important element of the EDI reform agenda.

Despite existing political will at the highest levels of government to include a public benefits agenda, the initial "Stage 1 Blueprint" did not provide for electrification and energy access programs until late in the document's development process. The reasons for this omission are not easily understandable. It might be that the PwC team and its expertise were oriented primarily toward the business and financial details of EDI consolidation. Another reason for the omission could be the government did not request a public benefits focus in its initial tender for EDI technical advisors to which PwC responded. One longtime observer of South Africa's reform process notes that:

"The public benefits issue was relegated to a secondary role. I am not sure that the PwC team itself had the available expertise to understand how rural electrification, in support of economic development, would be crucial to the success of the REDs model and the attainment of South Africa's RDP objectives."[22]

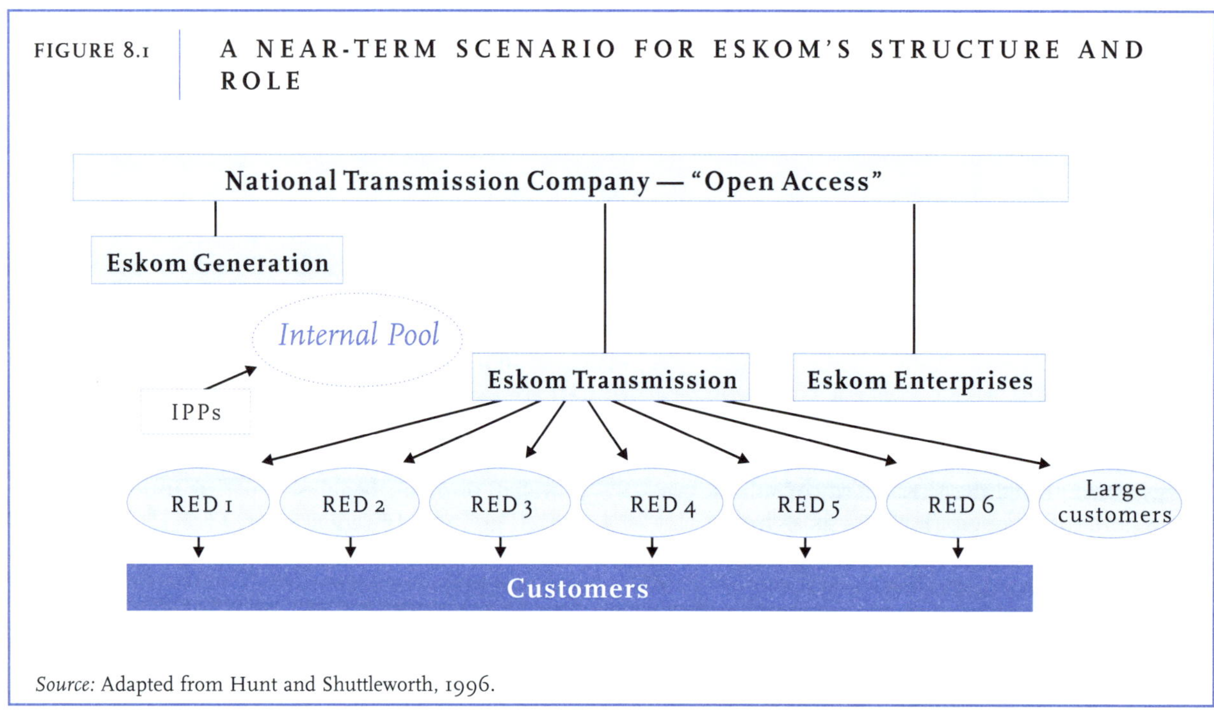

FIGURE 8.1 | A NEAR-TERM SCENARIO FOR ESKOM'S STRUCTURE AND ROLE

Source: Adapted from Hunt and Shuttleworth, 1996.

Local Governments

The South Africa Local Governments Association (SALGA), comprised of municipal governments, has not always supported consolidation into the six REDs. From their point of view, consolidation would cause the loss of financial and political leverage by the municipality. It would also infringe upon the status of SALGA's members as Service Authorities (as guaranteed by South Africa's constitution) to deliver electricity services. By the mid-1990s, the municipal distributors realized that most of them did not have the resources to deliver electricity services properly and they agreed that a mechanism for sharing resources was required. The agreement did not allay concerns about the costs associated with the consolidation process.

The bottom line for many distributors is day-to-day financial survival.

For most distributors the bottom line is more about day-to-day survival and less about delivering public benefits programs to their customers (Clark, 2001).[23] Many argue that the basis upon which they are currently operating is too fragile and too uncertain to consider public benefits programs. When asked why no DSM programs existed in his municipality, one employee of the distribution business replied "you are asking me to run before I can walk!"[24] The direction that electricity sector reform follows ultimately will have significant impact on the future role of environmental benefits such as renewable energy and energy efficiency (Winkler and Mavhungu, 2001). Private investors will likely reduce investment in public-benefit energy efficiency, for example, as there is little incentive to invest in measures that reduce revenue (Clark and Mavhungu, 2000).

Box 8.4 summarizes some of the key actors in the EDI reform process and describes their specific interests and issues of concern.

The ESI Design Process and Public Benefits

Until now, most discussions have occurred on a high-level basis, generally between the DME and DPE ministers and Eskom executives. A wider range of supply-industry stakeholders has not yet participated in the ESI design process. The government has not yet provided a forum for public participation probably due to the nascent state of ESI reform plans, at least compared to EDI reform plans. The government is still developing its own policy in this area (Clark, 2001). To some other stakeholders it is not clear that Eskom is sufficiently "broken" to warrant a complete restructuring (Clark, 2001). The following provides a sample of different ESI stakeholder viewpoints and positions.

National Government

In Johannesburg in early 2000, a World Bank-sponsored workshop convened for the first time all the important government stakeholders required for a serious discussion about ESI reform.[25] Workshop participants focused primarily on mainstream reform issues, although some also called for attention to public benefits in their final statements. In particular, the DPE and DME ministers called for (1) electrification; (2) energy access; (3) energy efficiency and DSM; and (4) protection for research and development. DME Minister Mlambo-Ncguka consistently raised concerns about the broad goals of electricity sector reform. She expressed particular concern about the effect that reform might have on the rural electrification program and black economic empowerment. Not discussed was a mechanism to support public benefits simultaneously with a move toward a competitive, privatized market.

Eskom has resisted reform plans, asserting the utility is well-equipped to deliver public benefits.

On May 30, 2001, the government demonstrated its desire to have some partial private domestic and some foreign equity investment in Eskom's genera-

Actors	Interests/Issues
End users	
Customers	Low electricity prices, improved services, reliable, and high quality power supplies.
Labor/COSATU	Low electricity prices, no job losses, competitive wages/salaries, a national distribution company.
Environmental Advocacy Groups	Limited environmental impact, sustainable development.
Distribution industry	
Municipal Electricity Distributors	Favorable asset transfer and shareholding (control) in new REDs structure, limited job losses, municipal levy, continued Service Authority status.
Eskom Executive Management	Continued natural monopoly status, no impact on Eskom credit rating, shareholding in distribution companies.
South Africa Local Government Association (SALGA)	Smooth transition, municipal levy, fair stake in RED structure.
Government	
Department of Minerals and Energy (DME)	Low electricity prices, financially viable distribution industry, ongoing electrification programs and other programs aimed at achieving rural development objectives.
Department of Public Enterprises (DPE)	Maximizing returns from Eskom distribution shares.
Department of Finance/National Treasury (DOF)	Financially viable distribution industry, transparent fiscal impact.
National Electricity Regulator (NER)	Financially viable distribution industry that is easier to regulate, rationalized tariff structures.

Source: Clark, 2001.

tion business. In a move to restructure Eskom, the Parliamentary Portfolio Committee on Public Enterprises instituted the Eskom Conversion Act. This legal process transformed Eskom into a public company—termed "corporatization"—in which the State is the sole shareholder. In early 2002, DPE announced plans to make 10 percent of Eskom's generation capacity available to black economic empowerment actors. DPE also stated that the government would sell an additional 20 percent to a strategic equity partner to secure foreign direct investment in Eskom (*Business Day*, 2002). Eskom will no longer be the sole shareholder. According to news reports, Eskom will retain a "dominant" role in the electricity generation industry for the foreseeable future. Dominant means that Eskom, a public sector enterprise, holds 70 percent of the generation business, and thus retains primary ownership and operational control.

<table>
<tr><td style="vertical-align:top; width:25%;">BOX 8.5</td><td style="vertical-align:top;">INTERESTS AND ISSUES OF SELECTED ESI REFORM STAKEHOLDERS</td></tr>
</table>

Actors	Interests/Issues
End users	
Customers	Low electricity prices.
Labor/COSATU	No privatisation, no job losses, competitive wages.
Environmental Advocacy Groups	Clean generation of electricity, positive social and environmental impact.
Supply industry	
Eskom Executive Management/Electricity Council	Limited reform, transmission to be transferred to Eskom Enterprises, limited competition.
Prospective and Existing Independent Power Producers (IPPs)	Fair access to transmission system, cost reflective tariffs, transparent regulatory framework.
South Africa Local Government Association (SALGA)	Unknown – not yet consulted.
Government	
Department of Public Enterprises (DPE)	Restructuring of state-owned assets to maximize returns to the state.
Department of Minerals and Energy (DME)	Efficient and competitive electricity supply industry, electrification program, opportunities for black economic empowerment.
Department of Finance/National Treasury (DOF)	Efficient and competitive electricity supply industry.
National Electricity Regulator (NER)	Efficient and competitive electricity supply industry, opportunity to develop new regulatory framework.

Source: Clark, 2001.

Eskom Management

Eskom management has been very vocal about ESI reforms, and its statements have contributed to the government's indecisive stance on this issue (Clark, 2001). Eskom has resisted electricity sector reform plans asserting that it is well-equipped to deliver public benefits. For instance, the utility has noted its own success with electrification, and raised the concern that the goal of electrification may not be as well-served in a restructured sector. These assertions contribute to a continuing government concern that moves to introduce competition into the ESI could have seriously detrimental effects on black economic empowerment and electrification. Similarly, Eskom has argued in the past against issuing licenses to new independent power producers (IPPs) on the basis that the utility itself can meet additional electricity demand through its DSM program.

Box 8.5 summarizes some of the key actors in the ESI reform process and describes their specific interests and issues of concern.

CONCLUSION

South Africa's choice to move slowly and test a more
market-oriented approach for its electricity sector is
rooted in its commitment to determine for itself what
is in the nation's best interest. The country's approach to electricity sector reform is "homegrown"—
rather than a product of financial pressures and
political influences from other governments, international investors, donors and development agencies,
and foreign consultants—with a preference for
competition and privatization.[26] As evidence,
consider South Africa's stance toward the World
Bank. Unlike many other countries in Africa and
around the world, the World Bank's involvement in
South Africa's reform process has been and will
likely continue to be minimal. Its only role to date
has been to provide funds for research and workshops. South Africa's attitude is likely the product of
a reactive response by South African stakeholders to
previous negative experiences with the Bank within
both the country and elsewhere in Africa (Clark,
2001).[27] At the same time, the World Bank displays a
tendency to suggest privatization as the solution to
sector management problems. It tends not to explore
an array of alternatives to revise and adapt current
practices (Steyn, 1995). This tendency likely adds to
South Africa's attitude toward the Bank.

The government's rationale for a more market-
oriented approach appears based more on its aver-
sion to financial risk in the electricity sector than on
influences and pressures from such international
financial institutions as the World Bank (Galen,
1998). The risks it would rather avoid are twofold.
First, in the case of a state-owned utility such as
Eskom, the investor is the state. Therefore, government alone is the bearer of the financial risk associated with an investment of billions of Rand in new
electricity generation capacity. Second, the amount of
investment needed over the next 20 years is substantial. Accordingly, it represents a significant degree of
financial risk to the investor. The government cannot
afford to be the sole investor. A critical decision the

government faces is whether Eskom should build the
next power plant or encourage competing investors
such as IPPs to enter the market (Eberhard, 1999).

South Africa's social and economic development
priorities—and its self-assessment of the need for
electricity sector reform—bode well for a broad
public benefits agenda. At the same time, the reform
process is nascent in South Africa in comparison to
many other countries. The country's experience with
its foreign consultants and the "Stage 1 Blueprint"
reflects that despite political will, public benefits
advocates still have to convince some key stakeholders of the link between social and environmental
concerns and the need to implement concrete
programs to address those concerns. It is too soon to
evaluate whether and how the reform process has
served individual public benefits so far—particularly
environmental benefits. Yet, there is tangible
progress toward the inclusion of such benefits in the
reform process.

South Africa's reform experience suggests a reason
to affirm the centrality of good governance—in the
form of implementing institutions with well-defined
roles, the rule of law and effective regulation, a stable
and enabling market structure, and transparent,
participatory public policy at the national and local
levels—to the future success of a public benefits
agenda. South Africa is grappling with concerns and
questions about public benefits early in the EDI and
ESI reform processes. The country has taken steps to
address stakeholder questions in policy statements
and to reflect stakeholder concerns in local governance structures. The DME minister has been vocal
and effective in bringing social concerns—particularly concerns about access to electricity—on to the
reform agenda, with the DPE minister also playing a
constructive role. The cabinet's intervention in the
"Stage 1 Blueprint" resulted in an additional emphasis on various public benefits programs. The intervention illustrates political will translated into action in
South Africa. These are hopeful signs that future
sector reform in South Africa will have a human face.

NOTES

1. This chapter draws on Alix Clark, 2001, "Power Sector Reform and the Public Benefits Imperative: A South African Case Study." Unpublished paper, Energy Development Research Centre, Capetown, South Africa (July). Online at http://www.wri.org/governance/iffe/powercases.html. This paper was written as part of a collaborative project on power sector reform and public benefits in developing and transition economies coordinated by the World Resources Institute.

2. Apartheid was a policy that called for the separation of people on the basis of race, codified into law.

3. Eskom is the fourth largest electric utility in the world, generates 95 percent of South Africa's electricity, and produces half of all the electricity consumed on the African continent (NER, 1999).

4. The RDP is a 150-page document outlining the ANC's macroeconomic strategy, including its plans to create jobs through public works; build a million new houses with electricity and plumbing; extend primary health care; provide 10 years of free access to education and redistribute land.

5. This tariff prompted some large, industrial customers to call for market liberalization, which would increase the pressure to remove such subsidies (Galen, 2002).

6. The *White Paper on the Energy Policy of the Republic of South Africa 1998* is the product of widespread public participation in the formulation of South African energy policy.

7. The government has been discussing instituting a poverty tariff and providing between 20 to 60 kilowatt-hours of free electricity per month to approximately 1.4 million poor households.

8. Meaning that generation, transmission, and distribution—the basic functions of a central-station, public utility—are not subject to competition in a marketplace for their services.

9. This includes 3,556 megawatts of mothballed capacity; and 1,429 megawatts of capacity under construction.

10. The RDP remains as the government's statement on South Africa's social and economic development goals.

11. The four SOEs are: Eskom (electricity utility); Denel (arms); Transnet (transport); and Telkom (telecommunications).

12. COSATU's press release is online at: http://www.cosatu.org.za/press/2001/eskom_Conversion_Bill (June 11, 2002).

13. An important point to consider is that corporatization is not the same as, and does not necessarily lead to, privatization. Our conjecture on Eskom's position is that it does want to attract additional equity by becoming a company with shares to sell. It is less likely, however, that Eskom wants to lose operational control to private investors or the market.

14. In 1993 the National Electricity Forum (NELF), a multi-stakeholder group, began to review electricity sector reform issues and practices, including an exploration of ESI reform issues.

15. Personal communication with Michael Eckhart, Solar International Management, Inc., Washington, D.C. (April, 2002).

16. South Africa's low electricity prices do not take into account external costs.

17. In the medium- to long term, more capacity will inevitably translate into higher prices. In the future, price considerations may become an important ESI reform driver.

18. Government is in the process, however, of developing a White Paper on Renewable Energy due in spring 2002.

19. South Africa will host the WSSD in September 2002.

20. The PwC developed the paper series under the Electricity Distribution Restructuring Project. The paper series is online at: http://www.dme.gov.za/energy/edi.htm (June 11, 2002).

21. The government did not publish the final document known as the "Stage 1 Blueprint," a government report on EDI reform. The PwC "Working Paper 7: Consolidated Emerging Views" served as the public document for stakeholder scrutiny and comment. Much of the content of "Working Paper 7" formed the basis of the government's "Stage 1 Blueprint." Cabinet adopted the PwC's recommended model of EDI reform in the final, revised "Stage 1 Blueprint" which reflected the electrification program's importance in the reform process.

22. Quote from a U.S.-based expert on electricity sector reform and rural electrification, April 2, 2002. All interviews for this chapter were conducted on a not-for-attribution basis. Consequently, interviewees are identified only by their institutional affiliation.

23. Personal communication with Michael Eckhart, Solar International Management, Inc., Washington, D.C. (April, 2002).

24. Interview with staff of a distribution utility in South Africa, 2000.

25. The workshop was attended by the DPE Minister, the DME Minister and Deputy Minister, NER Executive Management, Eskom Executive Management, and senior staff of various government departments including DME, DPE, and Finance (now called Treasury). Union officials were invited but, apart from a South African Mining Workers' Union representative, did not attend. Invited were international electricity sector experts to give presentations on the experiences of Argentina, Australia, Columbia, Chile, France, Hungary, New Zealand, Norway, and the United Kingdom.

26. Personal communication with Michael Eckhart, Solar International Management, Inc., Washington, D.C. (April, 2002).

27. In its South Africa Country Assistance Strategy, the World Bank itself noted the following:
"Since re-engaging in South Africa in the early 1990s, the Bank has played an active role. At the outset, the Bank had a strongly negative image, particularly among ANC cadres who viewed the Bank through the lens of their experience in other African countries undergoing structural adjustment. The Bank responded by adapting its focus...and pursuing an inclusive dialogue with all segments of society, inside and outside of government. Establishment of a more productive relationship with government and other groups has improved the perception of the Bank in South Africa, although distrust and ambivalence about the Bank's motives and agenda persist with certain groups." Online at: http://www.worldbank.org (June 10, 2002).

REFERENCES

Chalmers, Robyn. 2002. "Who Will Come to Power Party?" *Business Day.* (March 1).

Clark, Alix. 2001. "Power Sector Reform and the Public Benefits Imperative: A South African Case Study." Unpublished manuscript. Energy and Development Research Centre. University of Cape Town, Cape Town, South Africa.

Clark, Alix and J. Mavhungu. 2000. "Promoting Public Benefit Energy Efficiency Investment in New Power Contexts in South Africa." Energy and Development Research Centre. Cape Town: The University of Cape Town.

Department of Minerals and Energy (DME). 1998. *White Paper on the Energy Policy of the Republic of South Africa 1998.* Pretoria. Online at: http://www.dme.gov.za/ publications/wp_ene/whitepaper1998.htm# Ministerialforward (June 10, 2002).

———. 2000. *Renewable Energy Strategy 2000.* Pretoria. Online at: http://www.dme.gov.za/energy/ restrategy_part3.htm (June 10, 2002).

Department of Public Enterprises (DPE). 2000. *An Accelerated Agenda Towards the Restructuring of State Owned Enterprises.* Hatfield: Republic of South Africa.

Eberhard, A. 1999. "Competitive Jolt Good for Consumers." *Business Day.* (November 11):15.

———. 2001. "Planning the Future of Eskom." *Business Day.* (October 24). Online at: http://www.bday.co.za/ bday/content/direct/1,3523,953756-6078-0,00.html (May 20, 2002).

Galen, P. 1998. "Grappling with Change: The South African Electricity Supply Industry." Report No. NREL/ TP-260-25454. Golden, Colorado: The U.S. National Renewable Energy Laboratory.

Hunt, S. and G. Shuttleworth. 1996. *Competition and Choice in Electricity.* New York: Wiley and Sons.

Mlambo-Ngcuka, P. 2001. Speech presented to the Africa Energy Forum: Power and Gas 21, Lyons, France, June 24-26, 2001. Online At: http://www.dme.gov.za/ newscentre/speeches/speecheshistory (June 10, 2002).

———. 2002. Parliamentary media briefing, Pretoria, South Africa. Online at: http://www.dme.gov.za/ speeches (February 15, 2002).

National Electricity Regulator (NER). 1999. *Statistical Overview of the Electricity Supply Industry of South Africa 1998.* Sandton, South Africa. Online at: http:// www.ner.org.za/stats.htm (May 3, 2002).

———. 2000. "Competition in Electricity Markets." *Electricity Regulatory Journal.* (September): 1-6. Online at: http://www.ner.org.za/publications/erj/sep2000.pdf (May 3, 2002).

PricewaterhouseCoopers (PwC). 2000. "Working Paper 7: Consolidated Emerging Views." Electricity Distribution Industry Restructuring Project. Online at: http://www.dme.gov.za/energy/ Stage%20I%20emerging%20views.pdf (June 10, 2002).

Radebe, J. 2000. "Restructuring for Development: Ensuring Benefits for the People." Address presented to the Privatisation Africa 2000 Conference, Sandton, South Africa. (May). Online at: http://www.dpe.gov.za/docs/ sp/2000/sp0516.html (June 10, 2002).

Steyn, G. 1995. "Restructuring the South African Electricity Supply Industry: Appropriate Governance in a Newly Democratized South Africa." *Utilities Policy*, Vol. 5 (2):95-108.

Wamukonya, N. 2001. "The Uneven Road for the Non-grid Programme in South Africa." *Energy for Sustainable Development.* Vol. 5 (3):41-46.

Winkler, H. and J. Mavhungu. 2001. "Green Power, Public Benefits and Electricity Industry Restructuring." Final report prepared for the Sustainable Energy and Climate Change Partnership: A Project of Earthlife Africa Johannesburg and WWF Denmark. Cape Town: The University of Cape Town.

9

CONCLUSION

Navroz K. Dubash

The six country studies included in this study suggest the need for a reframing of electricity reform in much of the developing world and in transition economies. Thus far, reform processes have left little space for articulation and promotion of a sustainable development agenda. The sector is being transformed to operate along commercial lines, which will likely bring some benefits. However, longer-term environmental and social interests in the management and operation of the sector get little attention from technocrats. Electricity reforms to date suggest a need for concern, and for action, if public benefits are not to be swept aside in the rush toward a market-driven vision of the electric power sector.

This conclusion draws on the country cases to identify the factors that have motivated reforms; discuss how reforms have been shaped by the politics in the sector; examine the role that donor agencies have played in those politics; and detail the extent to which public benefits concerns have been addressed in the reform process. The chapter concludes with a discussion of the way forward toward a politics of reform that incorporates a public benefits agenda.

THE DRIVERS OF ELECTRICITY REFORM

The case studies suggest that electricity reform was overwhelmingly driven by the need to establish financial health in the sector. This framing of the objectives around financial health, rather than around, for example, the provision of electricity services in an environmentally sustainable manner, proved central to the conceptualization and implementation of reforms.

Several countries undertook reform against the backdrop of severe macroeconomic crises, which contributed to the emphasis on financial health. Argentina was grappling with a heavy debt burden and saw privatization of its debt-laden utilities as one way out. Bulgaria was faced with massive inflation and a declining standard of living. The East Asian crisis plunged the dominant Indonesian public utility into bankruptcy, forcing consideration of a sell-off of assets to raise capital. And in Ghana, a recommendation to privatize the distribution utility was part and parcel of a decade of structural adjustment of the Ghanaian economy. Even where reforms did not take place in the context of a systemic macroeconomic crisis, staunching the sector's drain on public finances was at the center of a larger financial reform package, as was the case with state-level reforms in India.

The case studies confirm the view that power sector reforms are part of a broader process of economic liberalization and integration into the global economy. Reforms were often driven by an immediate need for capital to maintain existing generation capacity and develop new capacity. To attract these funds, governments had to increase space for the private sector and create an environment that would provide adequate returns to private capital. Donor agencies played an important role in reinforcing this trend. The initiation of reforms in India, Indonesia, Ghana, and Bulgaria was directly tied to a withdrawal of international donor support for the power sector in these countries.

However, raising money on international capital markets has been a considerable challenge for most developing and transition economies. With technologically run-down and institutionally flawed distribution networks, India and Ghana were not very attractive to private capital, except under the generous terms provided to independent power producers (IPPs). Even in Argentina, where the public utility had established a respectable track record, privatization was accomplished at fire-sale prices. Developing and transition economies were left with a conundrum—private investment was unlikely to be forthcoming without an assurance of adequate returns, yet to increase returns many countries needed an infusion of capital. Faced with this situation, concerns over finance loomed large, and colored the entire process of reforming the power sector.

Power sector reforms are part of a broader process of economic liberalization and integration into the global economy.

The only partial exception to this trend is the South African experience. While attracting foreign investment has been important in South Africa, the country has not undertaken reforms under the cloud of a macroeconomic crisis, the public utility is in relatively good health, and while finances are an issue, they are not as pressing as in some of the other countries. Indeed, the additional political context for reforms—the post-apartheid transition and concerns of black empowerment—has created some space for discussion of reforms that support the social goals of equity and empowerment. Most important, it has enabled an explicit political commitment to social issues—in particular, enhancing access to electricity—as central to reforms.

Using power sector reforms to attract capital has not had a sucessful track record. To attract private

capital, governments have paid a high price—financially unviable and politically undesirable deal-sweeteners. Moreover, a decline in available private financing since 1997 calls into question whether, even with generous incentives, the electricity sector is sufficiently attractive for the private sector. Structuring reforms only to attract finance may not be a sustainable strategy.

THE POLITICS OF ELECTRICITY REFORM

With financial motivations the primary driver of reform, other interests have not been well represented, and the politics of reform in the countries studied have been largely exclusionary. There has been little political space in reform debates to define and articulate a public benefits perspective. Yet, reform processes have not been written on a blank slate. The legacy of past decisions, and latent but potentially powerful political interests have also played a role in defining reform. This section examines each of these themes in turn, and concludes with a discussion of whether and how a more open architecture could benefit the process of electricity reform.

The Actors Who Shaped Reform, and Those Who Were Excluded

That the goal of reforms was narrowly defined as financial led to an orientation weighted to technocrats. Moreover, a concern that interest groups could potentially hinder the smooth progress of reform left little scope for the inclusion of outside voices.

In Argentina, a homegrown set of bureaucrats laid out the ground rules for reform along strict economic lines, and worked to implement this vision to the exclusion of social factors. In India, international aid agencies played an important role in initiating reform and in emphasizing privatization as a principal goal. The details were worked out by international consultants with technical experience in economic and institutional issues. In Bulgaria, a reconstituted state regulatory agency initially controlled the reform process, ignoring pressures from national actors and international donors. This situation was reversed only with a change in government. In Ghana, control over reforms rested with domestic bureaucrats, but Chilean consultants provided key design insights. State technocrats maintained a tight focus on the problem definition—institutional restructuring to ensure financial health.

In some cases, the intervention of politicians broadened the scope and goals of the reform process. In South Africa, the active intervention of the Minister of Mines and Energy and the Minister of Public Enterprises was responsible for highlighting concerns regarding access to electricity in reform debates. In Indonesia, active oversight by the Minister of Mines and Energy not only kept the process on track while he was in power, but also led to engagement with stakeholders through a series of regular "breakfast meetings." In Bulgaria, the Parliament, assisted by the World Bank, raised concerns that a proposed energy law included an unjustified supply orientation and insufficiently independent institutions. These few episodes suggest that the involvement of politicians, with a more direct finger on the national pulse, may contribute to active consideration of public concerns.

However, the reform processes in the country studies were as notable for which groups were absent as for those that were present. The processes were shaped by technocrats (and occasionally politicians) from energy agencies and finance ministries. Representatives with line authority for environment or rural development were almost totally excluded. It is startling that in six case studies, there was not a single mention of a ministry of environment getting a seat at the table. This absence is a further reflection of the narrow focus of power sector reforms, and the perception that reforms are a technical matter with little scope for broader debate.

Equally, if not more startling, is that few nongovernmental organization (NGO) voices emerged to participate actively in the design of reforms, despite the

vibrant civil society in many countries studied here, and the worldwide growth in nongovernmental organizations. In Argentina, there was a complete absence of civil society input during the reform process. In India and Indonesia, NGO activists focused almost exclusively on criticism of IPP projects and the IPP policy, with only one or two groups in each country engaged in the broader reform process. In Bulgaria, NGOs had limited success in injecting a focus on energy efficiency into reforms, until a change in political conditions created new opportunities. While international NGOs have been extremely active and influential for over a decade in international environmental negotiations, few have made the link between national reform processes and global environmental outcomes. The few exceptions—such as the efforts of Greenpeace to promote renewable energy in Argentina, an analysis of reforms in Indonesia by the Washington, D.C.-based NGO Bank Information Center, and a technical scenario analysis of the impacts of reform on climate change by the Pew Center on Global Climate Change—have not been central to national reform processes.

This limited role of environmental or rural development interests—governmental or nongovernmental—suggests that power sector technocrats have successfully reinforced the perception that reforms are a technical issue, and that social and environmental concerns do not belong at the table. The constrained space for engagement has restricted efforts to change this perception. Moreover, the limited capabilities and skills available among advocacy groups to express their concerns in the technical debates over reforms, or for that matter among socially and environmentally focused government agencies, is an obstacle to broadening the debate.

Finally, few private sector entities—whether representing traditional technologies or promoting renewable energy technologies—played a direct role in shaping reforms. The one partial exception is that of the Independent Power Producers Association of India as a convening organization in the early stages of reform. This lack of influence by corporate actors contrasts with a direct role by energy companies, such as the part played by the Enron Corporation in the United States, in shaping reforms to support their comparative advantage (Wayne, 2000). The lack of a similar dynamic in the cases studied here may be because there are few national corporations positioned to benefit from reforms, and international corporate actors have been inadequately placed to shape national processes.

Political Constraints in the Reform Process

Technocrats have structured reforms as an exclusive process because of a perception that the sector has been prey to political capture in the past. And, indeed, the case studies offer some evidence to support this perception. This section describes past political pressures in the sector and the congealed institutional interests around past policies. It examines how both have constrained current reform efforts.

In all the countries studied, the difficulty of raising or restructuring tariffs has proven to be the single biggest challenge to restoring the sector to ensure its financial health. Past flawed policies to provide preferential prices to vote blocs have created well-mobilized interest groups, which can be a considerable political obstacle to tariff changes. These groups are not the ones most in need of concessional prices but are comparatively well off—such as large farmers in India and middle-class urban consumers.

Market reformers have argued that withdrawing subsidies from relatively well-off populations would be socially just and would help the financial condition of the sector. They have urged politicians to carry out such reforms. Even if political obstacles were overcome, however, there are additional and very real public interest concerns about raising prices. In

Ghana, the real cost of electricity is amplified by a falling currency. Rising electricity prices form an alarmingly large share of the budget of one vulnerable group—retirees—in Bulgaria. Electricity priced at the cost of supply—as market reforms aim to do—would price many low-income groups out of the market. In a reflection of these pressures, tariff increases have led to popular uprisings in Argentina, India, Indonesia, Ghana, and South Africa. Recognizing that regressive tariffs were socially unsustainable, a subsequent set of reforms in Argentina included a more equitable restructuring of electricity tariffs. Since financial motivations are central to reform, the politics of tariff increases—which include the vested interests that obstruct them and the vulnerable communities that are in need of support—are central to the success of any reform approach.

That the electric power sector provided considerable scope for political patronage is well supported by the case studies. Evidence of corruption in the allocation of contracts in Indonesia; the use of a large Energy Resource Fund free of public oversight to allocate funds to political allies in Bulgaria; and manipulation of prices in India to serve vote blocs all demonstrate the entanglement of the sector in self-serving politics. The costs in terms of efficiency and the public interest were considerable.

Political manipulations within the sector have created institutional interests around past policies, which in turn have been obstacles to reform. In India, politicians are reaping the bitter fruit of past populist policies, as farmers successfully oppose reversal of a long-standing policy of concessional or free electricity for agriculture. In Ghana, a long-term contract for power with a giant aluminum smelter has been a sticking point for reforms.

A decade of preferential concessions to Independent Power Producers (IPPs), particularly through over-generous power purchase agreements, is one of the most damaging legacies of past policies. In India, IPPs became a route to attract private capital without addressing the politically thorny institutional problems at the heart of the sector, and raised concerns over the process by which contracts were negotiated. In Indonesia, the introduction of 27 IPPs provided fertile ground for the growth of corruption and graft. In both countries, flawed IPP projects led to a broader cynicism about the role of the private sector. Subsequent privatization efforts may well be contaminated by the taint of this decade-long debacle. IPPs have also cast a shadow over the economics of reform. Financial sweeteners provided to attract IPPs locked public utilities into unviable contractual obligations, which have proved an obstacle to more fundamental reforms, and in particular to privatization. While these problems were most manifest in Asia, the push toward increasing capacity through IPPs in Bulgaria and Ghana suffered from some of the same flaws.

Existing public utilities have exercised a considerable drag on reform processes in several countries. In these cases, it is difficult to separate bureaucratic interests from often credible arguments that the death knell has been prematurely sounded for the public utilities model, despite a track record of success. In Ghana, the Volta River Authority (VRA) has strenuously fought efforts at integration into a post-reform sector, arguing that it has been a profitable utility for four decades. VRA's critics have countered that profits are based on its privileged position as a quasi-enclave, and that the general populace—much of which has no access to electricity—has benefited little from VRA's profitable status. By contrast, while South Africa's dominant utility, Eskom, has also benefited from some preferential treatment, it can point to a credible track record of expanding access to electricity to rural areas. Whether or not their arguments were valid, efforts to overcome resistance from existing utilities and the bureaucracies that run them have influenced the reform process.

Since public utilities have traditionally been large employers, labor unions have been a potent political force in opposing reforms. In Indonesia, the public utility's union challenged reforms as unconstitutional. In Bulgaria, reforms were met with an organized campaign of opposition. While the utilities may indeed be inefficiently over-staffed—as in India, where employment in the sector is a means of patronage—labor concerns are a legitimate issue. In both Argentina and Bolivia, labor unions won a share in the equity of privatized state enterprises, demonstrating the potential for political compromise. In South Korea, labor unions began a promising dialogue with environmentalists. Although they failed to come up with a shared platform, the two groups managed to substantially narrow their differences. Whether through the need for accommodation, or through the creation of new alliances, organized labor will continue to shape the political context for reform.

Open versus Closed Political Process?

The introduction to this study discussed the relative merits of a quick and stealthy approach to reform versus one based on an open process aimed at building a social consensus. This section revisits this discussion in light of the country studies.

In most countries, there was limited public consultation built into the reform process. Argentina, the earliest reformer among those considered here, is also the one that was most exclusive. In Indonesia, the Asian Development Bank sought to establish a consultation process with civil society groups. However, this effort faced obstacles, both because of NGO reluctance to participate in a donor funded program, and because NGOs felt the consultation would occur only after many decisions had already been made. A similar concern was reported about the consultation effort in India. While bureaucrats and donors pointed to a good-faith effort at engaging consumer organizations and NGOs during state-level reforms, those consulted felt that they were being informed of decisions after the fact in an effort to win their support, rather than having their opinions solicited. Only in South Africa have reforms arguably gone through a broad-based consultation process, with stakeholders invited to submit their views at the early goal-defining stage. The narrow range of actors involved, combined with limited consultation, suggest that most countries have followed the quick and stealthy approach to reforms.

Since reforms are still underway in many of the countries studied here, the information required to evaluate the stealthy versus the consultative approach is necessarily incomplete. In Argentina, the earliest reformer of those studied, a closed process certainly resulted in rapid implementation of a reform blueprint. However, Argentina later had to revisit several decisions in an effort to address social issues; this suggests that the closed approach failed to incorporate significant public concerns. In India, a closed approach led to a cycle of institutional reform in one state, Orissa, but the benefits and political viability of the model are in doubt. In Indonesia, Bulgaria, and Ghana, reform processes have been slow to develop and have been bogged down in local political processes. South Africa provides a partial example of a more open approach, but one that is in a preliminary stage. Perhaps most intriguing is the California experience, where a broad range of stakeholders was deeply engaged in the details of the reform process. For some, the subsequent problems were due to unwise efforts to pander to special interests, while others suggest the failure was caused

by unrealistic assumptions at the outset. Hence, the country cases do not demonstrate the success of the closed approach, nor do they provide sufficient evidence to fully endorse a more open politics of reform.

While not conclusive, the cases do provide circumstantial evidence that building a social consensus around reforms may strengthen outcomes and their political viability. First, reforms in the power sector are unlikely to lead to immediate and demonstrable benefits to the public. In India, for example, reforms in the state of Orissa resulted in tariff increases, but few immediate improvements in service. In Argentina, service quality did improve, but benefits were weighted toward the well-off. For reforms to win long-term political support, a good case can be made that the polity must buy into a vision for the sector, which requires a consensus-driven approach.

Second, the need for political consensus is even more evident when tariff increases are an issue. In Ghana, increases by the government led to considerable protest and a rapid reversal of tariff hikes, while similar increases by a more credible independent agency won popular acceptance. To the extent that consultation is an important ingredient in credibility, this example reinforces the value of a more open process. Finally, an exclusive process aimed at avoiding capture by past vested interests is open to capture by the new wielders of authority. In Bulgaria, for example, the state agency charged with reform in the late 1990s took on an increasingly authoritarian bent, defying national constituencies and international agencies alike.

Whether out of belief in its benefits, or because policymakers are being subject to greater popular scrutiny, reform processes are increasingly open. Modifications to early reforms in Argentina were initiated in part by a wave of public dissatisfaction and because more attention was paid to the impacts of reform on the public. In India, a national electricity bill was circulated widely and debated more robustly than a similar measure prepared five years earlier during the initial round of state-level reforms. The latest reformer, South Africa, has demonstrated far more effort at open debate and consultation than have countries that began earlier. Finally, this trend is coincident with a growing normative consensus: open processes support democratic institutions, which over time are better able to represent the long-term interests of citizens.

THE ROLE OF DONOR AGENCIES

Donor agencies, led by the World Bank, have played a significant part in reform efforts in several countries. The case study experience suggests that donors can play a critical role in nudging countries toward making politically difficult decisions. At the same time, for donors to control the direction of reforms is inappropriate and undermines creation of a national consensus in support of reforms. The country examples also suggest that donors have made only half-hearted efforts at promoting attention to a public benefits agenda in the course of reforms, and have been hampered in their interventions in the sector by a past history that has weakened their reputation.

Multilateral donor agencies exercising their financial leverage have been central to initiating reforms in a number of countries. For example, the World Bank required India and Ghana to commit to institutional reform as a condition of loans for the power sector. IMF conditionality associated with a macroeconomic crisis jump-started the reform process in Bulgaria and restarted a stalled process in Indonesia. In contrast, reforms in Argentina and South Africa were largely homegrown.

It is relevant to revisit here the distinction made in the introduction between firmness in initiating a reform process compared to the need for consensus-building around consolidation of reforms. While donor agencies may have played an important role in cutting through a domestic political morass in initiating reforms, it is far less certain that a continued heavy hand in consolidating reforms was appropriate. In Ghana, the World Bank unsuccessfully sought to impose a management contract approach, one that it deemed most appropriate for African countries at the time. This approach drew

from the experience of neighboring Cote d'Ivoire, where the World Bank urged greater involvement for the private sector despite the lack of an appropriate regulatory framework. In Bulgaria, the IMF attempted to force a rigid adherence to tariff increases, without considering solutions to the resultant social costs and potential for political upheaval. In India, the World Bank implemented a model based on privatization of state utilities to stem fiscal losses. In contrast to donor intervention in Ghana and Bulgaria, intervention in India was accompanied by a skillful process of building borrower ownership for the approach advocated by the World Bank. This ownership was limited to state bureaucrats, however, and did not extend to the broader population. However skillful the input, the current imperfect state of knowledge of the viability of market-led reforms, and the importance of tailoring reform processes to local economic and political conditions, calls for a less-directive approach by donor agencies.

Conditionality has proved effective at initiating reforms, but a more subtle strategy of engagement with borrower countries has been more useful in the design and consolidation stages.

On occasion, donor agencies use their analytic capacity to make a case for how reforms can benefit the public. The World Bank deserves credit for supporting demand-side management components in India and Ghana, but this achievement is limited by the failure to examine the underlying institutional disincentives to DSM in a post-reform sector. In India, the World Bank conducted the only study of the environmental effects of restructuring, but under pressure from the Ministry of Power limited its scope to exclude questions central to the reform process, thereby restricting its utility. Other efforts were more productive. In Bulgaria, the World Bank's "Energy-Environment Review" made the case for greater efforts at end-use efficiency, challenging the government's self-serving orientation toward new sources of supply. This and similar studies proved useful when a new government sought to revisit the trajectory of reforms. In Ghana, the Danish government has been a keen supporter of efforts to promote renewable energy.

These examples suggest that while conditionality is the most obvious and blunt tool in the donor arsenal, it is not the only one. Conditionality has proved effective at initiating reforms, but a more subtle strategy of engagement with borrower countries has been more useful in the design and consolidation stages. In Indonesia, well-timed research was influential in shaping decisions about new generation capacity. In India, numerous analyses helped frame decisions about reform efforts. In South Africa, faced with considerable suspicion of donor intervention, the World Bank has limited itself to providing technical assistance on reform experiences in other countries, an approach that was acknowledged to be useful. To the extent that these efforts crystallized decisions and informed debates, they contributed to national processes of building a consensus.

Donor agencies' efforts are restricted by their weak reputation in many borrower countries. Due to a decade and more of structural adjustment policies, their intervention is associated with economic hardship. In Bulgaria, for example, IMF policies have been associated with harsh price increases. In Indonesia, the Asian Development Bank (ADB) funded a process of public awareness and engagement in reforms. The effort faltered due to NGO reluctance to work with the ADB, an institution they perceived not to be working in the interest of sustainable development in the country.

Suspicions about donors have been worsened by industrialized country efforts to promote the interests of their own corporations. For example, in Indonesia, one arm of the U.S. government sought to promote a large U.S.-funded IPP even as an advisor supported by its aid agency, USAID, cautioned against the project. In India, the U.S. government has interceded on behalf of the Enron Corporation on a disputed project, despite the strong reservations expressed by both internal critics and the World Bank.

The credibility dilemma is illustrated by donor engagement with countries regarding IPPs. In both India and Indonesia, the World Bank urged caution, and on occasion denounced particular projects in strong terms—an intervention that should reflect well on the institution's concern for the long-term health of the sector and for the public interest. At the same time, the World Bank was using its leverage to promote more space for the private sector in each country despite the lack of a suitable institutional framework, which aroused the ire of its critics. Thus, while donor institutions may intervene in positive ways on occasion, these interventions tend to be overshadowed by the public perception that they are pursuing their own agenda. In the case of the power sector, this agenda was directed at creating a sector that could sustain private sector profit-making and release the public sector from the burden of poorly functioning utilities. Civil society groups perceived the promotion of such an approach—without an explicit articulation of how reforms served the public interest—as not in the national interest.

WHAT ROLE FOR PUBLIC BENEFITS IN REFORMS?

The country studies suggest that, with the possible exception of South Africa, reforms were driven by attention to restoring financial health in the sector through institutional restructuring and greater involvement of the private sector. In South Africa, the government made a political commitment to promote social goals through reforms. Here, we summarize evidence to support the theme that while restoring financial health does automatically contribute to a measure of socially and environmentally beneficial outcomes, a *laissez-faire* approach by itself is insufficient to incorporate attention to public benefits. Instead, once reforms are "locked in," it is hard to retrofit them. For this reason, a political commitment to sustainable development outcomes is necessary if reforms are to actively promote public benefits.

Public Benefits as Secondary to Implementing Reforms

Evidence from all the case studies suggests that, for reform advocates, ensuring a financially viable sector was the most relevant definition of the public interest. Certainly this viewpoint was the explicit starting point of studies conducted in Bulgaria and India prior to reforms. Of the cases studied here, only the Argentine experience provides sufficient information to comment on the social and environmental outcomes of reforms.

In Argentina, while the quality of service did improve on average, in many other ways the social outcomes were largely negative. The reform process led to a decline in access to electricity for low-income urban groups, as aggressive measures were taken to disconnect illegal consumers, with no mitigating steps to soften the social impact. Efforts to connect the 10 percent of Argentina's population in rural areas without access to electricity services have been largely unsuccessful. Moreover, changes in electricity prices have had a regressive impact. While prices fell overall, large consumers enjoyed declines of up to 71 percent, while prices charged to lower-income groups remained essentially unchanged.

With regard to environmental benefits, reform advocates believed that a more efficiently run sector was likely to be less environmentally burdensome. The evidence from Argentina partially substantiates this view. For example, transmission and distribution losses fell from 27 percent to below 10 percent in the five years after reform was initiated. In addition, the promotion of natural gas plants over older thermal power plants led to a net increase in the efficiency of generation. At the same time, the new organization of the sector considerably eroded incentives for distributors to invest in energy efficiency. Efforts to invest in renewable energy sources were largely unsuccessful, in part because of poor political support.

The Argentine experience provides strong evidence that social benefits, particularly equity goals, are not automatically served by reform processes. On environmental issues, the evidence reinforces the discussion in Chapter 2, which suggested that reforms may automatically lead to short-term gains, but may not signal a more sustainable long-term energy future.

The Problem of "Lock-in"

While achieving financial stability undoubtedly contributes to the public interest, the country studies suggest that this narrow approach also excludes consideration of a broader definition of public benefits. This assessment has focused on access to electricity services, considerations of equity, and promotion of environmental sustainability. To the extent that these concerns were embraced in the countries studied, they were seen as separable from, sequential to, and of lower priority than a successful restructuring and privatization of the sector.

A laissez-faire approach is insufficient to incorporate attention to public benefits.

In Argentina, for example, the first round of tariff restructuring was driven entirely by theoretical economic criteria—such as requiring that prices reflect the cost of supply—even though such rules often led to inequitable outcomes. Only much later did the Argentine government and its donors agree that some deviation from market principles, including the possible use of cross-subsidies, was necessary to ensure the public interest. In Ghana, despite low levels of access to electricity and an existing publicly funded effort to expand the grid, only late in the reform process was there any effort to bring private participation together with a program of increasing access built around public funds. In Bulgaria, despite the country's agreement to greenhouse gas emission targets under the Kyoto Protocol, there was little consideration of this commitment while restructuring a sector that emitted 56 percent of the country's carbon emissions.

The case studies offer good reasons to doubt the viability of a sequential view of reforms and public benefits—one that argues for fixing the financial position of the sector first and dealing with public benefits later. Specifically, technical, political, and institutional decisions are "locked in" so that current decisions constrain future choices. This rigidity can make it hard to retrofit the sector to address public benefits once the framework for reforms is set.

The IPP experience in Asia offers a good example of both technological and institutional lock-in. In both India and Indonesia, IPPs were invited to provide new supply capacity, with no effort at assessing available supply options and their full costs. As a result, the sector was locked into technologies chosen within the existing institutional framework, which tended to favor large generating plants. In addition, since large IPPs typically locked public utilities into buying power whether it was needed or not, they were also a constraint on further institutional reform in the sector.

The creation of new regulatory institutions is an important example of institutional lock-in. The formal mandates of these regulators, the institutional culture of regulatory agencies, and their expertise are relevant to their approach to public benefits in the future. Although the evidence from the case studies is thin, the experience in India clearly suggests that the enabling legislative framework does not empower regulators to internalize environmental considerations. This mandate is further constricted by international economic consultants, who train regulatory commissions to view their role in narrow terms as managing the tradeoff between profitability and short-term consumer interests. As a result, regulators approach their jobs based on the strict separability of economic and environmental decisions. Moreover, regulators are trained to work

toward short-term goals. Once enshrined, these views are hard to change.

Finally, since reform is often the result of intense negotiation among several interests, compromises that are made early in the process can constrain the opportunities for future outcomes resulting in political lock-in. In the process of reforms, deals were struck with existing utilities in South Africa and Ghana, and with municipalities in South Africa. In India, arrangements with farmers are a likely component of future reforms. If environmental and social interests are not at the table when these compromises are made, it may be difficult to renegotiate them at a later date without jeopardizing a fragile agreement. In Argentina, for example, an attempt at restructuring distribution tariffs in the interests of equity has proved hard because of pressure to maintain the sanctity of deals struck with private utilities at the time of privatization.

Social and Environmental Benefits

The case studies provide a scattering of examples in which reforms have led to explicit efforts to embrace a broader definition of public benefits. Where such approaches have been followed, they have been based on financial motivations or a broader political imperative.

Efforts at promoting more efficient use of energy on the demand side were common to several countries. Reformers facing severe financial constraints were attracted to the view that greater efficiency often more than pays for itself. The enormous scope for energy efficiency uncovered by the crisis in California, where energy is used far more efficiently than in most developing countries, illustrates the potential of this approach. India, Ghana, and Bulgaria have all included attention to end-use energy efficiency programs as part of reforms, and the issue is an important item in the South African reform debates. In Ghana, the idea that demand-side management could lower the impact of tariff hikes on the population and thereby dampen unrest was an important motivator. Similarly, in India, efforts to introduce more efficient use of electricity for agriculture were based on expectations that if farmers could achieve the same ends with less electricity, they would be less opposed to price increases.

Given the enormous potential benefits of end-use efficiency, substantially more focused attention to this avenue seems warranted. Since tariff increases are arguably the biggest stumbling block to the political sustainability of reforms, countries would benefit considerably from a sustained public campaign to increase efficiency of use across all consumer sectors. The programs in place do not demonstrate such a focus. Early efforts in India were limited by an inappropriate context, and also by limited political will from government bureaucrats and donors. In Bulgaria, where the scope for energy efficiency is enormous, progress has been slow until recently because greater end-use efficiency would have undercut the planned program of supply expansion supported by the then-political leadership.

Finally, while isolated programs toward energy efficiency have been put in place, there has been little effort to understand how unbundling can present challenges to promoting demand-side management—largely due to increased transactions costs— and to taking ameliorative measures. Only in South Africa has the debate been joined in consideration of incentives and disincentives for end-use efficiency in reform design processes.

While expanding access to electricity services has long been a focus of public policy in several countries prior to reforms, the country studies show that this concern was seldom integrated into utility reform. There was little consideration of how privatized utilities would contribute to greater access. Instead, public-led efforts at grid expansion typically continued in parallel to sector reforms, as in Ghana and India. The lack of systematic efforts to integrate increased electricity access with utility reform is a significant failure of donors and governments.

The shift away from public monopoly did hold out the possibility for involvement by a broader range of market participants, and a few efforts to meet access needs through creative decentralized solutions. In Ghana, a program that, in part, was intended to stimulate short-term supply using diesel generators quickly became an opportunity to promote decentralized renewable energy systems connected to the grid. During the second phase of reforms in Argentina, the government instituted a program to establish subsidized concessions for electricity supply to dispersed populations; these would be auctioned based on the lowest subsidy bid. One component of the program funded by the Global Environment Facility and the World Bank aimed at using renewable energy technologies to meet these needs. Despite a promising start, the project has languished for want of political support. A similar effort in Chile to attract commercial participation in rural electrification has had more success. This program has considerably expanded access, on occasion using off-grid renewable energy technologies. Finally, Morocco's successful experience with a cooperative program between the state utility, municipalities, and the private sector suggests that public monopolies can also develop creative programs for grid expansion.

Donor agencies have often been in the lead in promoting programs tied to public benefits objectives. Examples include the role of Danish aid in promoting renewable energy in Ghana; Global Environment Facility support for Argentina's rural renewable energy program; and USAID's support for energy efficiency in India. Yet, these programs have often been marginal to the main thrust of reforms, and have not been sufficiently knit into a broader articulation of the public interest. Without domestic political support for the objectives that these programs represent, they are prey to changes in national political sentiment and trends in donor assistance.

In all the case study countries, the shift to a more decentralized, market-led approach has brought no broad vision for future development of the sector, nor the corresponding accountability for developing such a vision. As the cases suggest, a post-reform sector will not substitute for such a vision. Problems left inadequately addressed include several concerns related to the success of the reforms and the long-term public interest. Most significant is the absence of a strategy with which to manage the tariff problem—balancing the need to raise revenues against the need to mitigate harm to vulnerable populations. Similarly, neither the potential costs of global efforts to address climate change—such as pressures to limit country emissions—nor opportunities, such as possible new sources of finance, are factored into current reform efforts.

TOWARD A PROGRESSIVE POLITICS OF ELECTRICITY REFORM

Integrating environmental and social benefits into power sector reform in developing and transition economies will continue to be a daunting challenge. Not only is reform technically complex, but the combination of macroeconomic crisis, entrenched political interests, and the centrality of finances crowd out attention to environmental and social factors. Indeed, from among the cases discussed here, there are few encouraging stories, and few successes on which to draw.

However, the country studies do offer insights into how reform is currently shaped, and therefore into how attention to the public interest can be reinserted into these processes. If there is one general lesson, it is that if public benefits are to receive attention, they must be shifted closer to the mainstream of political concerns that drive reform. There are several steps to achieving this objective.

1. Frame reform around the goals to be achieved in the sector.

The first imperative is to widen the definition of the problem that power sector reform is attempting to solve. A narrow focus on institutional restructuring driven by financial concerns is too restrictive to promote a sustainable development agenda. Admittedly, in a crisis context, there is little time, scope, or patience for a wide-ranging discussion on public priorities. For this reason, the initiation of discussion over reform and its goals must occur well prior to the onset of a crisis. The most appropriate framing for inclusion of a public benefits agenda would begin with an articulation of a future vision of the sector consistent with the services that a reformed sector is intended to provide. For example, if expanding access to electricity is a national priority, it is necessary to ensure that reforms promote and do not undermine progress toward this goal. While donor agencies often play a central role in initiating reform, they must step back during the process of defining goals to allow a nationally driven vision of reform to emerge. This vision will likely vary to fit national circumstances. To encourage a more visionary approach to reforms in the electric power sector:

Governments can:

- reframe the goals of reform to encompass social and environmental concerns and highlight sustainable development objectives;
- broaden participation within government to provide a seat at the table to ministries charged with promoting environmental protection, rural development, and poverty alleviation.

Donor agencies can:

- continue to provide funds conditional on evidence of reform, but only if accompanied by domestic ownership over the form and content of those reforms;
- help governments to conduct analyses of the scope for inclusion of public benefits concerns in a reform process, recognizing that there is greater space for such analyses when governments are not facing crisis situations.

- develop a vision of a post-reform sector that promotes sustainability, access to electricity, and equity in pricing;
- undertake a campaign of public outreach to build support for a public benefits-oriented reform process.

2. Structure finance around reform goals, rather than reform goals around finance.

Reform processes thus far have been propelled by a drive to attract capital, and reform goals have catered to this end. Since protection of the public interest can require tradeoffs with opportunities for profit, reform processes must move beyond the imperative of attracting capital at the expense of promoting other objectives. While this may seem a far-fetched notion in capital-constrained developing countries, the time may now be opportune to change the terms on which private capital enters a country.

Over the last decade, governments have attracted capital by providing favorable terms and risk guarantees. Privatization, particularly of distribution, has been based on the promise of tariff increases to ensure a profit. These measures have not won popular backing, and, as a result, have not been politically sustainable. The net result is that politically illegitimate guarantees of low risk and high return have not been socially sanctioned, and have proved ephemeral. Indeed, overall trends in private finance to developing countries have gone down, likely in response to increased perceptions of risk. A broader vision of reform and a public consensus supporting that vision could lower these risks. Private capital may be willing to accept more realistic financial returns, if they are combined with less risk. Political legitimacy in a reform program, tied to some innovation in mechanisms for raising finance, may be a more promising route than tailoring reforms to short-term profit horizons.

Moreover, some approaches to public benefits, such as private promotion of decentralized power sources for rural areas, can be financed through innovative financing models. Finally, it would be useful to explore other avenues to bring additional resources to the sector to promote a public benefits agenda. The scope for exploiting flexible trading mechanisms under the Kyoto Protocol offers one possible avenue.

While the argument above provides an alternative route to obtaining capital, there is no doubt that maintaining financial viability in the sector will continue to be a problem. Advocates who promote public benefits are likely to receive a more sympathetic hearing if they present their arguments in financial terms. Specifically, it is necessary to confront head-on the argument that revenues are inadequate to meet costs in many developing and transition economies, and that any efforts at a public benefits agenda that require bearing a financial burden, while desirable, are unaffordable. Once again, end-use efficiency provides an important example of a cost-saving approach that also provides social and environmental benefits. To encourage a shift in this direction:

Governments can:

- seek to attract capital based on the lower risk associated with a socially sanctioned and popular reform program.

Donor agencies can:

- reconsider their advocacy of reform designed to attract private capital using contractual risk management instruments, in favor of reform designed to attract private capital on the basis of risk management through sound governance;
- help developing countries to complement existing financing by mobilizing alternative sources of finance to support sustainable development, including mitigation of global environmental problems.

Civil society can:

- conduct analysis that demonstrates the technical and financial scope for incorporation of a public benefits agenda;

- build international coalitions to influence the policies of external actors relevant to national power sector reforms, particularly multilateral donor agencies.

Private sector actors can:
- direct investments to countries that signal a social consensus around power sector reforms and a corresponding diminution of risk;
- develop innovative financing models to support national objectives for reform.

3. Support reform processes with a system of sound governance.

An open-ended framing of reform will only reflect public concerns if it is supported by a robust process of debate and discussion. Hence, a third imperative is to embed debate over power sector reforms in a sound process of governance guided by transparency, openness, and participation. Such an approach is more likely to provide the political space for articulation of a range of public concerns than has the closed process prevalent thus far. It is also more likely to build public consensus in support of reforms, making for a more politically sustainable process.

Technocrats who design reforms are correctly fearful that open and transparent processes could also allow space for vested interests to undermine these reforms. While this is a danger, it is balanced by the likelihood that a more open process can help build broader political support. While no single case study provides a convincing example of good governance leading to inclusion of public benefits, there is circumstantial evidence to support this link. For example, in India, the emergence of some open and transparent regulatory agencies has spurred development of civil society capacity and action to participate effectively in regulatory decisions. There is also evidence to suggest that lack of good governance undermines public benefits. The opaque process by which IPPs were welcomed into Asia, leading to the erosion of public interest and confidence, is a case in point. To promote good governance in the sector:

Governments can:
- design an open process of goal definition for reform that includes space for meaningful public consultation and input early in the reform process;
- ensure that a post-reform sector builds in mechanisms for public feedback, consultation, and adjustment;
- create a legal framework for the independent operation of electricity regulators supported by openness to information and consultation.

Donor agencies can:
- include requirements that decisionmaking be based on practices of good governance;
- support capacity building in civil society to enable participation in reform decisions, and to play an oversight role in the governance of the sector.

Civil society can:
- build advocacy strategies around a call for good governance in the process of policy reform, to complement project advocacy;
- establish a long-term capacity to continually monitor and engage with regulatory institutions, which would ensure continued attention to public benefits in a post-reform sector.

4. Build political strategies to support attention to a public benefits agenda.

Finally, it is important that advocates of a sustainable development agenda strengthen political coalitions in support of public benefits and counter those supporting parochial interests. In particular, the case studies suggest that social concerns carry far more political weight in a national context than do either local or international environmental issues. Efforts to exploit links between social and environmental agendas would likely be a useful political approach.

For example, end-use energy efficiency promises both environmental and social gains, and is politically resonant since it blunts the impact of tariff increases. A campaign to place efforts at end-use efficiency at the center of reform could, therefore, win wide support from social advocates, environmentalists, consumer groups, and industrialists. Advocating a focus on access to electricity through environmentally sound technologies offers another example, albeit a less politically potent one. Since rural populations are hard to mobilize, the scope to build a broad-based coalition for this issue is limited. Nonetheless, rural development practitioners, environmentalists, and emergent industries in the renewable energy arena could usefully join hands in support of this approach. To achieve these outcomes:

Governments can:

- emphasize attention to social and environmental benefits as part of the mandate, capabilities, and culture of regulatory agencies, within a framework of goal-setting at the political level.

Donor agencies can:

- be accountable for analyzing their own lending operations for institutional reform to ensure a minimum "do-no-harm" standard, and to identify opportunities to proactively promote public benefits;
- incorporate attention to public benefits as part of their advisory and technical work, including the use of consultant expertise.

Civil society can:

- develop national capacity to conduct advocacy around both the politics and the technical dimensions of reform processes;
- use sound analysis to build national coalitions for promotion of particular components of a public benefits agenda, drawing on an understanding of local politics;
- use international coalitions to draw links with issues of global concern, particularly global climate change;
- develop coalitions with private sector actors with an interest in sustainable energy futures.

Private sector actors can:

- advocate policies that enable renewable energy technologies and energy efficiency measures to compete on a level footing with other supply options;
- build coalitions with NGOs and other supporters of sustainable energy futures to win political support for these policies.

By focusing on financial health, reform in the electric power sector has excluded a range of broader concerns also relevant to the public interest. In this study, we have focused on the social and environmental concerns at stake in electricity reform. We have found that not only are they inadequately addressed, but that socially and environmentally undesirable trajectories can be "locked-in" through technological, institutional, and financial decisions made now that constrain future choices. Consequently, social and environmental benefits need to be internalized in reform-related decisionmaking.

To do so, the process by which reform goals are defined, who participates in the reform process and decisionmaking must change to embrace a more consensus-driven design. More complex processes bring with them greater risks of capture by special interests and failure due to a cacophony of voices. Yet, non-inclusive reforms of the electricity sector these reforms have not incorporated the breadth of interests that deserve a voice and have not yet shown themselves to be sustainable—financially, socially, or environmentally. This study has suggested several reasons to believe that a modified approach guided by a vision of a socially and environmentally sustainable electricity future may yield a more satisfying outcome.

REFERENCE

Wayne, Leslie. 2002. "Enron, Preaching Deregulation, Worked the Statehouse Circuit." *New York Times.* February 9.

DANIEL BOUILLE is Vice President and Senior Researcher at the Institute for Energy Economics at the Bariloche Foundation in Argentina, and an advisor to the Environmental Directorate at the Argentine Foreign Affairs Ministry. He is also a professor at the University of Buenos Aires, the Argentine Catholic University, and is involved with other degree granting programs. He has published on energy planning, sustainable energy policies, greenhouse mitigation, and other environment and development related issues. He was the lead author of the Third Assessment Report of the Intergovernmental Panel on Climate Change working group on stabilizing greenhouse gases. Mr. Bouille received his postgraduate degree in Energy Economics from the University of Cologne, Germany.

ALIX CLARK is an independent researcher on energy issues. Her work focuses on public energy research and outreach and specifically on improving energy services for poor people. She coordinates a project on power sector reform and the public benefits imperative for the International Energy Initiative. She was previously programme leader for the markets and governance programme at the Energy Development Research Centre in Capetown, South Africa. She holds a masters degree in Economics from the University of Capetown.

DIMITAR DOUKOV is a Financial Analyst at the Center for Energy Efficiency—a non-governmental organization established in 1992. He was previously a Research Associate at the National Center for Regional Development & Housing Policy, Program Coordinator with ICMA/USAID and Head of Department at the Bulgarian Foreign Investment Agency. Mr. Doukov holds a Master's Degree in Economics and Management of Industry from the University for National and World Economy, Sofia, Bulgaria.

NAVROZ K. DUBASH is a Senior Associate in the Institutions and Governance Program at the World Resources Institute. His work explores the impact of financial globalization on problems of environment and development as part of WRI's International Financial Flows and the Environment (IFFE) project. He also works on mechanisms for democratic global governance, and on local institutions for natural resource management. Dr. Dubash holds Ph.D. and M.A. degrees in Energy and Resources from the University of California, Berkeley.

HILDA DUBROVSKY is a Researcher at the Institute for Energy Economics at the Bariloche Foundation in Argentina. She is a Professor in the Economics of Electricity, and has participated in technical assistance projects related to the economic, regulatory, institutional and environmental aspects of the energy sector. She brings an international perspective to her work, both through her academic training and her experience coordinating the Electric Integration of South America Project (European Union and CIER). She holds a postgraduate degree in Economy and Energy Planning.

ISHMAEL EDJEKUMHENE is a Senior Projects Officer of the Kumasi Institute of Technology and environment. His research interests focus on power sector reform, particularly the socio-economic impact of electricity privatization; policy options for removing barriers to renewable energy technologies; and the Clean Development Mechanism. He has a Masters in Economics (Public Policy) from the University of Hull, UK.

CRESCENCIA MAURER is a Senior Associate in the Institutions and Governance Program at the World Resources Institute. Her areas of research include financial flows and the environment, public participation in environmental decision-making, and urban environmental problems in Latin America. Ms. Maurer holds a master's degree in environmental management from Duke University.

ELENA PETKOVA is a Senior Associate in the Institutions and Governance Program at the World Resources Institute. Her work focuses on opening the political space for independent contribution to policy and public access to information and participation in environmental decision-making. She directs the "Access Initiative" a collaborative project on environmental governance, and has also worked on climate change issues, decentralization of natural resource management and local environmental action planning in Central and Eastern Europe. She has a Ph.D. in the History of International Relations.

JULIA A. PHILPOTT is a Senior Associate in the Climate, Energy, and Pollution Program at the World Resources Institute. Her research explores the implications of energy-related policies, private investment, and governance structures on sustainable development in developing countries. She also analyzes the global climate implications of private investment patterns in electricity and transport sectors worldwide. She holds a Master of Science in Urban and Regional Planning from the University of Wisconsin, Madison.

SUDHIR CHELLA RAJAN is a Senior Scientist at Tellus Institute. His work focuses on the interactions among social, political and technological issues relating to sustainable energy. He has a research background in meteorology and air quality modeling, transportation and mobile sources, the technical and environmental characteristics of energy technologies, rural energy systems, and power sector reform. Dr. Rajan holds a Ph.D. in Environmental Science and Engineering from the University of California, Los Angeles.

AGUS SARI is Executive Director of Pelangi, a non-profit, independent environmental think tank based in Jakarta, Indonesia. He is also a member of the Electricity Restructuring Working Group, a roster member of the Asian Development Bank Inspection Panel, a member of the Scientific Steering Committee of the Institutional Dimension on Global Environmental Change Project of the International Human Dimension Program, and an Advisor to the Indonesian Delegation to Climate Change Negotiations and the Preparatory Committee for the World Summit on Sustainable Development. He obtained his post-graduate degree in Energy and Resources from the University of California, Berkeley.

FRANCES SEYMOUR is the Director of the Institutions and Governance Program at the World Resources Institute. She directs projects on international financial flows and the environment, and sustainable development challenges in Southeast Asia. She also currently guides the Access Initiative, a global collaboration to promote respect for environmental procedural rights. She serves on the board of the International NGO Forum on Indonesian Development as well as on advisory committees for Human Rights Watch—Asia and the University of North Carolina's University Center for International Studies. She holds a masters degree from the Woodrow Wilson School at Princeton University, and a B.S. in Zoology from the University of North Carolina at Chapel Hill.

World Resources Institute

The World Resources Institute is an environmental think tank that goes beyond research to create practical ways to protect the Earth and improve people's lives. Our mission is to move human society to live in ways that protect Earth's environment for current and future generations.

Our program meets global challenges by using knowledge to catalyze public and private action:

- To reverse damage to ecosystems. We protect the capacity of ecosystems to sustain life and prosperity.
- To expand participation in environmental decisions. We collaborate with partners worldwide to increase people's access to information and influence over decisions about natural resources.
- To avert dangerous climate change. We promote public and private action to ensure a safe climate and sound world economy.
- To increase prosperity while improving the environment. We challenge the private sector to grow by improving environmental and community well-being.

In all of its policy research and work with institutions, WRI tries to build bridges between ideas and action, meshing the insights of scientific research, economic and institutional analyses, and practical experience with the need for open and participatory decision-making.